T0176854

CAMBRIDGE TRACTS IN MATHEMATICS

General Editors

J. BERTOIN, B. BOLLOBÁS, W. FULTON, B. KRA,
I. MOERDIJK, C. PRAEGER, P. SARNAK, B. SIMON, B. TOTARO

CAMBRIDGE TRACTS IN MATHEMATICS

GENERAL EDITORS

J. BERTOIN, B. BOLLOBÁS, W. FULTON, B. KRA, I. MOERDIJK, C. PRAEGER,
P. SARNAK, B. SIMON, B. TOTARO

A complete list of books in the series can be found at www.cambridge.org/mathematics.
Recent titles include the following:

188. Modern Approaches to the Invariant Subspace Problem. By I. CHALENDAR and
     J. R. PARTINGTON
189. Nonlinear Perron–Frobenius Theory. By B. LEMMENS and R. NUSSBAUM
190. Jordan Structures in Geometry and Analysis. By C.-H. CHU
191. Malliavin Calculus for Lévy Processes and Infinite-Dimensional Brownian Motion.
     By H. OSSWALD
192. Normal Approximations with Malliavin Calculus. By I. NOURDIN and G. PECCATI
193. Distribution Modulo One and Diophantine Approximation. By Y. BUGEAUD
194. Mathematics of Two-Dimensional Turbulence. By S. KUKSIN and A. SHIRIKYAN
195. A Universal Construction for Groups Acting Freely on Real Trees. By I. CHISWELL and
     T. MÜLLER
196. The Theory of Hardy's Z-Function. By A. IVIĆ
197. Induced Representations of Locally Compact Groups. By E. KANIUTH and K. F. TAYLOR
198. Topics in Critical Point Theory. By K. PERERA and M. SCHECHTER
199. Combinatorics of Minuscule Representations. By R. M. GREEN
200. Singularities of the Minimal Model Program. By J. KOLLÁR
201. Coherence in Three-Dimensional Category Theory. By N. GURSKI
202. Canonical Ramsey Theory on Polish Spaces. By V. KANOVEI, M. SABOK, and J. ZAPLETAL
203. A Primer on the Dirichlet Space. By O. EL-FALLAH, K. KELLAY, J. MASHREGHI, and
     T. RANSFORD
204. Group Cohomology and Algebraic Cycles. By B. TOTARO
205. Ridge Functions. By A. PINKUS
206. Probability on Real Lie Algebras. By U. FRANZ and N. PRIVAULT
207. Auxiliary Polynomials in Number Theory. By D. MASSER
208. Representations of Elementary Abelian p-Groups and Vector Bundles. By D. J. BENSON
209. Non-homogeneous Random Walks. By M. MENSHIKOV, S. POPOV, and A. WADE
210. Fourier Integrals in Classical Analysis (Second Edition). By C. D. SOGGE
211. Eigenvalues, Multiplicities and Graphs. By C. R. JOHNSON and C. M. SAIAGO
212. Applications of Diophantine Approximation to Integral Points and Transcendence.
     By P. CORVAJA and U. ZANNIER
213. Variations on a Theme of Borel. By S. WEINBERGER
214. The Mathieu Groups. By A. A. IVANOV
215. Slenderness I: Foundations. By R. DIMITRIC
216. Justification Logic. By S. ARTEMOV and M. FITTING
217. Defocusing Nonlinear Schrödinger Equations. By B. DODSON
218. The Random Matrix Theory of the Classical Compact Groups. By E. S. MECKES
219. Operator Analysis. By J. AGLER, J. E. MCCARTHY, and N. J. YOUNG
220. Lectures on Contact 3-Manifolds, Holomorphic Curves and Intersection Theory.
     By C. WENDL
221. Matrix Positivity. By C. R. JOHNSON, R. L. SMITH, and M. J. TSATSOMEROS
222. Assouad Dimension and Fractal Geometry. By J. M. FRASER
223. Coarse Geometry of Topological Groups. By C. ROSENDAL
224. Attractors of Hamiltonian Nonlinear Partial Differential Equations. By A. KOMECH and
     E. KOPYLOVA
225. Noncommutative Function-Theoretic Operator Function and Applications. By J. A. BALL
     and V. BOLOTNIKOV
226. The Mordell Conjecture. By H. IKOMA, S. KAWAGUCHI and A. MORIWAKI
227. Transcendence and Linear Relations of 1-Periods. By A. HUBER and G. WÜSTHOLZ

The Mordell Conjecture
A Complete Proof from Diophantine Geometry

HIDEAKI IKOMA
Shitennoji University

SHU KAWAGUCHI
Doshisha University

ATSUSHI MORIWAKI
Kyoto University

CAMBRIDGE
UNIVERSITY PRESS

University Printing House, Cambridge CB2 8BS, United Kingdom

One Liberty Plaza, 20th Floor, New York, NY 10006, USA

477 Williamstown Road, Port Melbourne, VIC 3207, Australia

314-321, 3rd Floor, Plot 3, Splendor Forum, Jasola District Centre,
New Delhi - 110025, India

103 Penang Road, #05–06/07, Visioncrest Commercial, Singapore 238467

Cambridge University Press is part of the University of Cambridge.

It furthers the University's mission by disseminating knowledge in the pursuit of
education, learning, and research at the highest international levels of excellence.

www.cambridge.org
Information on this title: www.cambridge.org/9781108845953
DOI: 10.1017/9781108991445

First published 2022

Printed in the United Kingdom by TJ Books Limited, Padstow Cornwall

A catalogue record for this publication is available from the British Library.

ISBN 978-1-108-84595-3 Hardback

Contents

Preface   *page* vii

1   What Is the Mordell Conjecture (Faltings's Theorem)?   1

2   Some Basics of Algebraic Number Theory   6
2.1   Trace and Norm   6
2.2   Algebraic Integers and Discriminants   9
2.3   Ideals in the Ring of Integers   11
2.4   Lattices and Minkowski's Convex Body Theorem   14
2.5   Minkowski's Discriminant Theorem   17
2.6   Field Extension and Ramification Index   22

3   Theory of Heights   25
3.1   Absolute Values   25
3.2   Product Formula   27
3.3   Heights of Vectors and Points in Projective Space   29
3.4   Height Functions Associated to Line Bundles   33
3.5   Northcott's Finiteness Theorem   39
3.6   Introduction to Abelian Varieties   43
3.7   Height Functions on Abelian Varieties   52
3.8   Curves and Their Jacobians   59
3.9   The Mordell–Weil Theorem   67

4   Preliminaries for the Proof of Faltings's Theorem   73
4.1   Siegel's Lemma   73
4.2   Inequalities on Lengths and Heights of Polynomials   77

v

4.3   Regular Local Ring and Index   85
4.4   Roth's Lemma   88
4.5   Norms of Invertible Sheaves   95
4.6   Height of Norm   99
4.7   Local Eisenstein Theorem   111

5   The Proof of Faltings's Theorem   117
5.1   Keys for the Proof of Faltings's Theorem   117
5.2   Technical Settings for the Proofs of Theorem 5.4, Theorem 5.5,
      and Theorem 5.6   129
5.3   Existence of Small Section (the Proof of Theorem 5.4)   134
5.4   Upper Bound of the Index (the Proof of Theorem 5.5)   143
5.5   Lower Bound of the Index (the Proof of Theorem 5.6)   146
5.6   An Application to Fermat Curves   156

      References   160
      Notation   163
      Index of Symbols   166
      Index   167

Preface

This book originated from course notes for "Topics in Algebra" taught by Atsushi Moriwaki to senior undergraduate students and beginning graduate students at Kyoto University in 1996. Shu Kawaguchi, then a graduate student, attended the course.

The purpose of the course was to give a self-contained and detailed proof of the Mordell conjecture (Faltings's theorem) by following Vojta's and Bombieri's papers [5, 29], while touching on several important theorems and techniques from Diophantine geometry.

We have fully revised and expanded the course notes into this book, and some of the explicit and detailed computations presented here may be appearing in the literature for the first time. This book will also provide an introduction to Diophantine geometry.

We assume that the reader is familiar with basic concepts of algebraic geometry and has good knowledge of undergraduate algebra and analysis. Some basics of algebraic number theory are included in Chapter 2.

For the reader who is familiar with the basics of Diophantine geometry, and is interested only in the proof of Faltings's theorem, we suggest starting from Chapter 5 while referring to Chapter 4. Otherwise, we suggest starting with Chapters 2 and 3 while referring, if necessary, to books on algebraic geometry and algebraic number theory (e.g., [11, 23]), and then reading Chapter 5 while referring to Chapter 4.

1

What Is the Mordell Conjecture (Faltings's Theorem)?

Diophantine geometry is the field of mathematics that concerns integer solutions and rational solutions of polynomial equations. It is named after *Diophantus of Alexandria* from around the third century who wrote a series of books called *Arithmetica*. Diophantine geometry is one of the oldest fields of mathematics, and it continues to be a major field in number theory and arithmetic geometry. If integer solutions and rational solutions are put aside, then polynomial equations determine an algebraic variety. Since around the start of the twentieth century, algebro-geometric methods have played an important role in the study of Diophantine geometry.

In 1922, Mordell (Figure 1.1) made a surprising conjecture in a paper where he proved the so-called Mordell–Weil theorem for elliptic curves (see Theorem 3.42). This conjecture, called the Mordell conjecture before Faltings's proof appeared, states that the number of rational points is finite on any geometrically irreducible algebraic curve of genus at least 2 defined over a number field. It is not certain on what grounds Mordell made this conjecture, but it was audacious at the time, and attracted the attention of many mathematicians. While some partial results were obtained, the Mordell conjecture stood as an unclimbed mountain before the proof by Faltings. Thus, when Faltings (Figure 1.2) proved the Mordell conjecture in a paper published in 1983, the news was circulated around the globe with much enthusiasm. Faltings's proof was momentous, using sophisticated and profound theories of arithmetic geometry. He proved the Shafarevich conjecture, the Tate conjecture, and the Mordell conjecture concurrently, and he was awarded the Fields Medal in 1986. Nevertheless, first-year students at universities can understand the statement of the Mordell conjecture, except for the notion of genus.

Let $f(X, Y)$ be a two-variable polynomial with coefficients in a number field K (e.g., the field $\mathbb{Q}$ of rational numbers). We assume the following:

Figure 1.1 Louis J. Mordell.
Source: Archives of the Mathematisches Forschungsinstitut Oberwolfach.

Figure 1.2 Gerd Faltings.
Source: Archives of the Mathematisches Forschungsinstitut Oberwolfach.

1. $f(X,Y)$ is irreducible as a polynomial in $\mathbb{C}[X,Y]$. Namely, if
    $f(X,Y) = g(X,Y)h(X,Y)$ with $g(X,Y), h(X,Y) \in \mathbb{C}[X,Y]$, then $g(X,Y)$
    or $h(X,Y)$ is a constant.
2. The algebraic curve C defined by $f(X,Y) = 0$, extended to the projective
    plane, is smooth. In other words, let $F(X,Y,Z) \in \mathbb{C}[X,Y,Z]$ be the
    homogeneous polynomial with

$$F(X,Y,1) = f(X,Y) \quad \text{and} \quad \deg F(X,Y,Z) = \deg f(X,Y).$$

Then the only solution in $\mathbb{C}^3$ of

$$F(X,Y,Z) = (\partial F/\partial X)(X,Y,Z) = (\partial F/\partial Y)(X,Y,Z)$$
$$= (\partial F/\partial Z)(X,Y,Z) = 0$$

is $(0,0,0)$.

In this setting, the algebraic curve C has genus at least 2 if and only if the
degree of f is at least 4. Thus, the Mordell conjecture states that if the degree
of f is at least 4, then the number of points $(a,b) \in K^2$ with $f(a,b) = 0$
is finite. Here, the assumption that f is irreducible is essential. Indeed, for

any polynomial $h(X,Y) \in K[X,Y]$, we set $f(X,Y) = Xh(X,Y)$. Then $f(X,Y) = 0$ has infinitely many solutions $\{(0,b) \mid b \in K\}$. On the other hand, the assumption that C is smooth is not essential. This assumption is made only to avoid the notion of genus.

Let us look at some examples. For simplicity, we assume for the moment that K is the field $\mathbb{Q}$ of rational numbers.

The quadratic polynomial $f(X,Y) = X^2 + Y^2 - 1$ satisfies assumptions 1 and 2, and the set of all rational solutions of $f(X,Y) = 0$ (i.e., points $(a,b) \in \mathbb{Q}^2$ with $f(a,b) = 0$) is equal to

$$\left\{ \left(\frac{1-t^2}{1+t^2}, \frac{2t}{1+t^2} \right) \,\middle|\, t \in \mathbb{Q} \right\} \cup \{(-1,0)\}.$$

Indeed, we associate a point $(a,b) \neq (-1,0)$ with $f(a,b) = 0$ to the slope t of the line $Y = t(X+1)$ that passes $(-1,0)$ and (a,b). Then the set of rational solutions of $f(X,Y) = 0$ other than $(-1,0)$ is in bijective correspondence with $\mathbb{Q}$. In this case, there are infinitely many rational points on the curve C defined by $f(X,Y) = 0$.

Next, we consider the quadratic polynomial $f(X,Y) = X^2 + Y^2 + 1$. It satisfies assumptions 1 and 2, but there are no rational solutions of $f(X,Y) = 0$. In general, if $f(X,Y)$ is a quadratic polynomial satisfying assumptions 1 and 2, then either there are infinitely many rational solutions of $f(X,Y) = 0$ or there are none. In other words, either there are infinitely many rational points on the curve C defined by $f(X,Y) = 0$ or there are none.

What about cubic polynomials? First we consider the cubic polynomial $f(X,Y) = X^3 + Y^3 - 1$. It satisfies assumptions 1 and 2. According to Euler (the cubic case of Fermat's last theorem), there are exactly four rational solutions $(\pm 1, 0), (0, \pm 1)$ for $f(X,Y) = 0$.

Next, we consider the cubic polynomial $f(X,Y) = Y^2 - X^3 - 877X$. It is easy to see that $f(X,Y)$ satisfies assumptions 1 and 2, and $(0,0)$ is a rational solution of $f(X,Y) = 0$. On the other hand, it is difficult to find a rational solution of $f(X,Y) = 0$ other than $(0,0)$. Perhaps surprisingly, there are in fact infinitely many rational solutions of $f(X,Y) = 0$, and the x-coordinate of the next "simple" rational solution (to be precise, the x-coordinate of a rational point with the next smallest Weil height in Chapter 3) is

$$\left(\frac{612776083187947368101}{78841535860683900210} \right)^2,$$

a result due to Bremner and Cassels. In general, the algebraic curve C defined by a cubic polynomial satisfying assumptions 1 and 2, extended to the projective plane, is equipped with the structure of an abelian group. Thus,

Figure 1.3 A curve of genus 2.

the set of rational points on C (on the projective plane) is also equipped with the structure of an abelian group, and the Mordell–Weil theorem (see Theorem 3.42) states that this group is finitely generated. In summary, if $f(X, Y)$ is a cubic polynomial satisfying assumptions 1 and 2, then there may be infinitely many rational points on the curve C defined by $f(X, Y) = 0$, but they are finitely generated as an abelian group.

What about polynomials of degree 4 or higher? This is where the Mordell conjecture comes in. It states that under assumptions 1 and 2, there are only finitely many rational points on the curve defined by the polynomial.

For any polynomial $\phi(X)$ in X of degree 4 or higher with coefficients in $\mathbb{Q}$, we put $f(X, Y) = Y - \phi(X)$. Then $\{(a, \phi(a)) \mid a \in \mathbb{Q}\}$ are rational points on the curve C defined by $f(X, Y) = 0$, and the curve C is smooth (on the affine plane). This may look strange at first glance, but in fact C has a singular point at the point $(0 : 1 : 0)$ at infinity on the projective plane, so assumption 2 is not satisfied.

Let $f(X, Y)$ be a polynomial of degree d satisfying assumptions 1 and 2, and let C be the curve defined by $f(X, Y) = 0$ (extended on the projective plane). The *genus* of C, which we have not explained so far, is equal to the number $(d - 1)(d - 2)/2$. Thus, the genus is 0 if $d = 1, 2$, 1 if $d = 3$, and at least 3 if $d \geq 4$. The genus of the curve defined by $Y - \phi(X) = 0$ is 0. Now the meaning of the Mordell conjecture becomes clearer. It states that the distribution of rational points is determined by the genus of the curve (Figure 1.3). Further, the genus of a curve is a topological invariant, realized geometrically as the number of "holes" of the curve. Thus, the Mordell conjecture states that a topological invariant controls rational points.

With its generality and innovativeness, no one thought that the Mordell conjecture would be solved before the turn of the century. The solution by Faltings was a monumental achievement in twentieth-century mathematics. In this book, we will call the Mordell conjecture "Faltings's theorem."

Perhaps Faltings's success lifted a mental block associated with the Mordell conjecture. Subsequently, Vojta and Bombieri found a relatively elementary proof in line with classical Diophantine geometry [5, 29]. The purpose of the present book is to give a self-contained proof of Faltings's theorem by following [5, 29], giving detailed accounts for some of the computations. Because the proof uses important results and techniques from Diophantine geometry, such as the theory of heights, the Mordell–Weil theorem, Siegel's lemma, and Roth's lemma, this book also serves as an introduction to Diophantine geometry. In this book, the reader will find the names of many great mathematicians who have contributed to the advancement of the field. In some sense, the path to Faltings's theorem ran alongside the advancement of mathematics more generally.

Lastly, we remark on some recent developments. In [9], Faltings proved that, if a subvariety X of an abelian variety defined over a number field does not contain a translation of a positive dimensional abelian subvariety, then the number of rational points on X is finite. A smooth projective curve of genus at least 2 is regarded as a subvariety of an abelian variety via the Jacobian embedding, and it does not contain a translation of a positive dimensional abelian subvariety. Thus, this result, often called Faltings's big theorem, generalizes Faltings's theorem on the Mordell conjecture. In this direction, the next big challenge will certainly be Lang's conjecture: if (the analytification of) a smooth projective variety X defined over a number field is a hyperbolic manifold, then the number of rational points on X should be finite. A smooth projective curve of genus at least 2 is a hyperbolic manifold, and thus, Lang's conjecture is a generalization of the Mordell conjecture. Very recently, Lawrence and Venkatesh [16] gave another proof of Faltings's theorem based on a detailed analysis of the variation of p-adic Galois representations, which does not use abelian varieties. Still the proofs of the Mordell conjecture that are known so far do not directly use hyperbolicity, and thus, are not applicable to Lang's conjecture. For example, the proof by Vojta and Bombieri in this book does not use the geometry of hyperbolic manifolds directly but instead uses some properties that are derived from the assumption that the genus g of the curve be at least 2 (e.g., $g > \sqrt{g}$, ampleness of a canonical bundle, and an embedding into the Jacobian variety). New ideas are needed for a direct proof of the Mordell conjecture that contributes to Lang's conjecture.

2

Some Basics of Algebraic Number Theory

In this chapter, we explain some of the basics of algebraic number theory, which we will need in Chapter 3 to introduce the theory of heights and to give a proof of the Mordell–Weil theorem. For more details on the basics of algebraic number theory, we refer the reader to the first three chapters of [23], for example.

2.1 Trace and Norm

Let F be a field, and let L be a finite separable extension field of F of degree $n = [L : F]$. We begin by defining the trace and the norm of an element of L over F.

For $x \in L$, we consider the linear map over F given by the multiplication by x

$$L \to L, \quad \alpha \mapsto x\alpha. \tag{2.1}$$

The *trace* $\mathrm{Tr}_{L/F}(x)$ and the *norm* $\mathrm{Norm}_{L/F}(x)$ of x over F are respectively defined by the trace and the determinant of the above map. Explicitly, we take a basis $\{\alpha_1, \ldots, \alpha_n\}$ of L as an F-vector space, and we write $x\alpha_j = \sum_{i=1}^n c_{ij}\alpha_i$ ($c_{ij} \in F$). Then the trace and the norm of x are respectively given by $\mathrm{Tr}_{L/F}(x) = \sum_{i=1}^n c_{ii}$ and $\mathrm{Norm}_{L/F}(x) = \det(c_{ij})$. They are independent of the choice of a basis.

We give equivalent definitions of the trace and the norm of x. Let Ω be an algebraically closed field containing F, and we set

$$\mathrm{Hom}_F(L, \Omega) = \{\sigma \mid \sigma : L \hookrightarrow \Omega \text{ is an embedding of fields with } \sigma|_F = \mathrm{id}_F\}. \tag{2.2}$$

Since L is a separable extension field of F, we have $\#(\mathrm{Hom}_F(L, \Omega)) = [L : F]$.

Lemma 2.1 *For any $x \in L$,*

$$\mathrm{Tr}_{L/F}(x) = \sum_{\sigma \in \mathrm{Hom}_F(L,\Omega)} \sigma(x) \quad and \quad \mathrm{Norm}_{L/F}(x) = \prod_{\sigma \in \mathrm{Hom}_F(L,\Omega)} \sigma(x).$$

Proof Let $F(x)$ be the field generated by x over F, and we set $[F(x) : F] = m$ and $[L : F(x)] = d$. Then $n = md$, and $\{1, x, \ldots, x^{m-1}\}$ is a basis of $F(x)$ over F. Let $\{\beta_1, \ldots, \beta_d\}$ be a basis of L over $F(x)$. Then $\{\beta_i x^j\}_{1 \le i \le d, 0 \le j \le m-1}$ is a basis of L over F. We denote by

$$X^m - c_1 X^{m-1} + \cdots + (-1)^m c_m \qquad (c_1, \ldots, c_m \in F), \qquad (2.3)$$

the minimal polynomial of x over F. Then, with respect to the basis $\{1, x, \ldots, x^{m-1}\}$, the matrix representation G of the F-linear map $F(x) \to F(x)$, $\alpha \mapsto x\alpha$ is given by

$$G = \begin{pmatrix} 0 & 0 & \cdots & 0 & (-1)^{m-1}c_m \\ 1 & 0 & \cdots & 0 & (-1)^{m-2}c_{m-1} \\ 0 & 1 & \cdots & 0 & (-1)^{m-3}c_{m-2} \\ \vdots & \vdots & \ddots & \vdots & \vdots \\ 0 & 0 & \cdots & 1 & c_1 \end{pmatrix} \in M(m, m; F).$$

Thus, with respect to the basis

$$\left\{ \beta_1, \beta_1 x, \ldots, \beta_1 x^{m-1}, \ldots, \beta_d, \beta_d x, \ldots, \beta_d x^{m-1} \right\}$$

over F, the F-linear map (2.1) has the matrix representation

$$\begin{pmatrix} G & O & \cdots & O \\ O & G & \cdots & O \\ \vdots & \vdots & \ddots & \vdots \\ O & O & \cdots & G \end{pmatrix} \in M(md, md; F).$$

It follows that $\mathrm{Tr}_{L/F}(x) = dc_1$ and $\mathrm{Norm}_{L/F}(x) = c_m^d$.

On the other hand, let $x_1, \ldots, x_m$ be the conjugates of x over F. Then $x_1, \ldots, x_m$ are the roots of (2.3), and we have $\sum_{k=1}^m x_k = c_1$ and $\prod_{k=1}^m x_k = c_m$.

For $\tau \in \mathrm{Hom}_F(F(x), \Omega)$, we set

$$\mathrm{Hom}_F(L, \Omega)_\tau = \{\sigma \in \mathrm{Hom}_F(L, \Omega) \mid \sigma|_{F(x)} = \tau\}.$$

Then $\#(\mathrm{Hom}_F(L, \Omega)_\tau) = [L : F(x)] = d$, so

$$\sum_{\sigma \in \mathrm{Hom}_F(L, \Omega)} \sigma(x) = d \sum_{\tau \in \mathrm{Hom}_F(F(x), \Omega)} \tau(x) = d \sum_{k=1}^{m} x_k = dc_1 = \mathrm{Tr}_{L/F}(x),$$

$$\prod_{\sigma \in \mathrm{Hom}_F(L, \Omega)} \sigma(x) = \left(\prod_{\tau \in \mathrm{Hom}_F(F(x), \Omega)} \tau(x) \right)^d = \left(\prod_{k=1}^{m} x_k \right)^d = c_m^d$$

$$= \mathrm{Norm}_{L/F}(x).$$

Hence, we obtain the assertion.   □

Let A and B be integrally closed integral domains with fractional fields F and L, respectively. Further, we assume that $A \subseteq B$ and that B is integral over A. For example, if B is a finite A-module, then B is integral over A. Comparing the definitions of the trace and the norm with Lemma 2.1, we obtain the following:

Lemma 2.2 *If $x \in B$, then $\mathrm{Tr}_{L/F}(x), \mathrm{Norm}_{L/F}(x) \in A$. Further, if $x \neq 0$, then $\mathrm{Norm}_{L/F}(x)/x \in B$.*

Proof We put $\mathrm{Hom}_F(L, \Omega) = \{\sigma_1, \ldots, \sigma_n\}$. By the definitions of the trace and the norm, we have $\mathrm{Tr}_{L/F}(x), \mathrm{Norm}_{L/F}(x) \in F$. Since x is assumed to be integral over A, $\sigma_i(x)$ is also integral over A. It follows from Lemma 2.1 that $\mathrm{Tr}_{L/F}(x)$ and $\mathrm{Norm}_{L/F}(x)$ are integral over A. Since A is integrally closed in F, we have $\mathrm{Tr}_{L/F}(x), \mathrm{Norm}_{L/F}(x) \in A$.

To show the last assertion, we may assume that σ_1 is the inclusion map in Lemma 2.1. Then $\mathrm{Norm}_{L/F}(x)/x = \sigma_2(x) \cdots \sigma_n(x)$. Since $\sigma_i(x)$ is integral over A, so is $\sigma_2(x) \cdots \sigma_n(x)$. Since $\mathrm{Norm}_{L/F}(x)/x \in L$ and B is integrally closed, we have $\mathrm{Norm}_{L/F}(x)/x \in B$.   □

We regard L as an F-vector space, and we define a symmetric bilinear form $(\,,\,)_{\mathrm{Tr}_{L/F}}$ over F to be

$$(\,,\,)_{\mathrm{Tr}_{L/F}} : L \times L \to F, \quad (x, y) \mapsto \mathrm{Tr}_{L/F}(xy). \qquad (2.4)$$

We call $(\,,\,)_{\mathrm{Tr}_{L/F}}$ the *trace form*. Further, for $\alpha_1, \ldots, \alpha_n \in L$, we set

$$D(\alpha_1, \ldots, \alpha_n) = \det((\alpha_i, \alpha_j)_{\mathrm{Tr}_{L/F}}), \qquad (2.5)$$

where $\left((\alpha_i, \alpha_j)_{\mathrm{Tr}_{L/F}}\right)$ is the Gram matrix. We call $D(\alpha_1, \ldots, \alpha_n)$ the *discriminant* of $\alpha_1, \ldots, \alpha_n$. By Lemma 2.1, if we set

$$\mathrm{Hom}_F(L, \Omega) = \{\sigma_1, \ldots, \sigma_n\},$$

then $\mathrm{Tr}_{L/F}(\alpha_i \alpha_j) = \sum_{k=1}^{n} \sigma_k(\alpha_i)\sigma_k(\alpha_j)$. We set $\Delta = (\sigma_i(\alpha_j))$. Then

$$D(\alpha_1, \ldots, \alpha_n) = \det(\mathrm{Tr}_{L/F}(\alpha_i \alpha_j)) = \det({}^t\Delta \cdot \Delta) = \det(\Delta)^2. \qquad (2.6)$$

Lemma 2.3 *The trace form* $(\ ,\ )_{\mathrm{Tr}_{L/F}}$ *is nondegenerate. In particular, for* $\alpha_1, \ldots, \alpha_n \in L$, $\{\alpha_1, \ldots, \alpha_n\}$ *is a basis of* L *over* F *if and only if* $D(\alpha_1, \ldots, \alpha_n) \neq 0$.

Proof   Recall that $(\ ,\ )_{\mathrm{Tr}_{L/F}}$ is nondegenerate if the Gram matrix with respect to one (and hence any) basis of L over F is invertible.

We take θ with $L = F(\theta)$. Then $\{1, \theta, \ldots, \theta^{n-1}\}$ is a basis of L over F. We put $\sigma_1(\theta) = \theta_1, \ldots, \sigma_n(\theta) = \theta_n$. Then Equation (2.6) gives

$$D(1, \theta, \ldots, \theta^{n-1}) = \left(\det(\theta_i^{j-1})\right)^2 = \left(\prod_{1 \leq i < j \leq n} (\theta_i - \theta_j)\right)^2 \neq 0.$$

Hence, $(\ ,\ )_{\mathrm{Tr}_{L/F}}$ is nondegenerate. The second assertion follows from the first assertion. $\square$

2.2 Algebraic Integers and Discriminants

A *number field* is a finite extension of the field $\mathbb{Q}$ of rational numbers. In this book, we often denote a number field by K.

Let K be a number field. An element $x \in K$ is an *algebraic integer* if there are a positive integer m and integers $a_1, \ldots, a_m$ such that

$$x^m + a_1 x^{m-1} + \cdots + a_m = 0.$$

The set of all algebraic integers, denoted by O_K, is called the *ring of integers* of K. In other words, O_K is the integral closure of $\mathbb{Z}$ in K.

We remark that given an extension ring B of A, the set of all elements of B that are integral over A is a subring of B (see, e.g., [18, Theorem 9.1]). The case of $A = \mathbb{Z}$ and $B = K$ implies that O_K is indeed a ring.

In this section, we use the trace form of the previous section and show that O_K, regarded as a $\mathbb{Z}$-module, is free and has rank n.

First, we put together some properties of the trace and the norm in the previous section in the case of $K/\mathbb{Q}$. First, for $x \in K$, the trace $\mathrm{Tr}_{K/\mathbb{Q}}(x)$ and the norm $\mathrm{Norm}_{K/\mathbb{Q}}(x)$ are elements of $\mathbb{Q}$. By $K(\mathbb{C})$, we denote the set of all $\mathbb{C}$-*valued points* of K. Namely, we set

$$K(\mathbb{C}) = \{\sigma \mid \sigma \colon K \hookrightarrow \mathbb{C} \text{ is an embedding of fields}\}. \qquad (2.7)$$

(With the notation in the previous section, we have $K(\mathbb{C}) = \mathrm{Hom}_{\mathbb{Q}}(K, \mathbb{C})$). Then $\#(K(\mathbb{C})) = [K : \mathbb{Q}]$, and for $x \in K$, it follows from Lemma 2.1 that

$$\mathrm{Tr}_{K/\mathbb{Q}}(x) = \sum_{\sigma \in K(\mathbb{C})} \sigma(x), \quad \mathrm{Norm}_{K/\mathbb{Q}}(x) = \prod_{\sigma \in K(\mathbb{C})} \sigma(x).$$

Further, it follows from Lemma 2.2 that, if $x \in O_K$, then $\mathrm{Tr}_{K/\mathbb{Q}}(x)$ and $\mathrm{Norm}_{K/\mathbb{Q}}(x)$ belong to $\mathbb{Z}$.

Proposition 2.4 *Let K be a number field with $[K : \mathbb{Q}] = n$, and let O_K be the ring of integers of K. Then O_K is a free $\mathbb{Z}$-module of rank n.*

Proof We take a basis $\{\alpha_1, \dots, \alpha_n\}$ of K as a $\mathbb{Q}$-vector space. Let $m \neq 0$ be an integer such that $m\alpha_1, \dots, m\alpha_n \in O_K$. Replacing α_i by $m\alpha_i$, we may assume that $\alpha_1, \dots, \alpha_n \in O_K$.

Let $\{\beta_1, \dots, \beta_n\}$ be the dual basis of $\{\alpha_1, \dots, \alpha_n\}$ with respect to $(\ ,\ )_{\mathrm{Tr}_{K/\mathbb{Q}}}$. Then, for any $x \in O_K$, we have $x = (x, \alpha_1)_{\mathrm{Tr}_{K/\mathbb{Q}}}\beta_1 + \cdots + (x, \alpha_n)_{\mathrm{Tr}_{K/\mathbb{Q}}}\beta_n$. Here, Lemma 2.2 implies that $(x, \alpha_i)_{\mathrm{Tr}_{K/\mathbb{Q}}} = \mathrm{Tr}_{K/\mathbb{Q}}(x\alpha_i) \in \mathbb{Z}$. We obtain

$$O_K \subseteq \mathbb{Z}\beta_1 + \cdots + \mathbb{Z}\beta_n.$$

Thus, O_K is a $\mathbb{Z}$-submodule of the free $\mathbb{Z}$-module $\mathbb{Z}\beta_1 + \cdots + \mathbb{Z}\beta_n$.

Since $\mathbb{Z}$ is a principal ideal domain, O_K is a free $\mathbb{Z}$-module. Let $\{\omega_1, \dots, \omega_r\}$ be a free basis of O_K as a free $\mathbb{Z}$-module, then $\{\omega_1, \dots, \omega_r\}$ is a basis of K as a $\mathbb{Q}$-vector space. Thus, $r = [K : \mathbb{Q}] = n$. It follows that O_K is a free $\mathbb{Z}$-module of rank n. $\qquad\square$

A free basis $\{\omega_1, \dots, \omega_n\}$ of O_K as a free $\mathbb{Z}$-module is called an *integral basis* of O_K. We denote the discriminant $D(\omega_1, \dots, \omega_n)$ (see Equation (2.5)) with respect to an integral basis $\{\omega_1, \dots, \omega_n\}$ of O_K by

$$D_{K/\mathbb{Q}} = D(\omega_1, \dots, \omega_n),$$

and call it the *discriminant* of K over $\mathbb{Q}$.

We remark that $D_{K/\mathbb{Q}}$ is independent of the choice of an integral basis. Indeed, let $\{\omega_1', \dots, \omega_n'\}$ be an integral basis of O_K. We write $\omega_j' = \sum_{i=1}^{n} a_{ij}\omega_i$, and set $A = (a_{ij})$. Then $A \in \mathrm{GL}(n, \mathbb{Z})$, and we have

$$\left((\omega_i', \omega_j')_{\mathrm{Tr}_{K/\mathbb{Q}}}\right) = {}^t A \left((\omega_i, \omega_j)_{\mathrm{Tr}_{K/\mathbb{Q}}}\right) A.$$

Thus,

$$D(\omega_1', \dots, \omega_n') = \det\left((\omega_i', \omega_j')_{\mathrm{Tr}_{K/\mathbb{Q}}}\right) = \det\left((\omega_i, \omega_j)_{\mathrm{Tr}_{K/\mathbb{Q}}}\right) = D(\omega_1, \dots, \omega_n).$$

Figure 2.1 Richard Dedekind.
Source: Getty Images / Mondadori Portfolio / Contributor.

2.3 Ideals in the Ring of Integers

Let K be a number field and let O_K be the ring of integers of K. We set $[K : \mathbb{Q}] = n$.

Recall that O_K is defined as the integral closure of $\mathbb{Z}$ in K. Thus, O_K is integrally closed in K. It follows from Proposition 2.4 that O_K is a free $\mathbb{Z}$-module of rank n, so in particular, O_K is a Noetherian $\mathbb{Z}$-module. By regarding ideals of O_K as $\mathbb{Z}$-modules, we see that any ascending chain of ideals of O_K is stabilized, so O_K is a Noetherian ring.

In general, if B is a ring that is integral over A and if $P_1 \subseteq P_2$ are prime ideals of B such that $P_1 \cap A = P_2 \cap A$, then $P_1 = P_2$ (see [18, Theorem 9.3], for example). In particular, the case where $A = \mathbb{Z}$ and $B = O_K$ tells us that any nonzero prime ideal of O_K is a maximal ideal.

A domain A that is not a field is called a *Dedekind domain* (Figure 2.1) if the following properties are satisfied: (1) A is integrally closed in the field of fractions; (2) any nonzero prime ideal is maximal; and (3) A is a Noetherian ring.

The above arguments show that O_K is a Dedekind domain.

The prime decomposition of ideals for Dedekind domains (see, e.g., [18, Theorem 11.6] or [23, Chapter I, Theorem (3.3)]) gives the following theorem:

Theorem 2.5 (Prime Decomposition of Ideals) *Let I be an ideal of O_K such that $I \neq (0), I \neq O_K$. Then there are nonzero prime ideals $P_1, \ldots, P_s$ of O_K such that*

$$I = P_1 \cdots P_s.$$

Further, the decomposition is unique (up to reordering of $P_1, \ldots, P_s$).

Let $I \neq (0)$ be an ideal of O_K. Putting together the same prime ideals that appear in the prime decomposition of I, we obtain that there are distinct nonzero prime ideals $P_1, \ldots, P_r$ of O_K and positive integers $e_1, \ldots, e_r$ such that

$$I = P_1^{e_1} \cdots P_r^{e_r}. \tag{2.8}$$

Note that, if $I = O_K$, we let $r = 0$. Further, the expression in Equation (2.8) is unique up to reordering of $P_1, \ldots, P_r$.

The prime decomposition of ideals of O_K is extended to that of fractional ideals of K. Here an O_K-submodule $I \neq \{0\}$ of K is called a *fractional ideal* of K if there is a nonzero element x of O_K such that $xI \subseteq O_K$.

Let I and J be fractional ideals of K. Then the product of I and J is defined by

$$IJ = \left\{ \sum_{\text{finite sum}} a_i b_i \,\middle|\, a_i \in I, \, b_i \in J \right\}. \tag{2.9}$$

Further, for a fractional ideal I of K, we set

$$I^{-1} = \{x \in K \mid xI \subseteq O_K\}.$$

Using the existence and uniqueness of prime decomposition of ideals of O_K, we have the following:

Theorem 2.6 *Let K be a number field, and let O_K be the ring of integers of K. Then we have the following:*

1. *The set of all fractional ideals of K forms an abelian group with multiplication given by (2.9). The identity element is O_K. The inverse of a fractional ideal I is given by I^{-1}.*
2. *Let I be a fractional ideal. Then there exist distinct nonzero prime ideals $P_1, \ldots, P_r$ of O_K and integers $e_1, \ldots, e_r$ such that I is uniquely decomposed as $I = P_1^{e_1} \cdots P_r^{e_r}$.*

In general, a ring A is called a *discrete valuation ring* if A is a principal ideal domain that is not a field, and A has a unique maximal ideal.[1] We take a generator t of the maximal ideal of A. We call t a *local parameter* of A. Then[2] $\bigcap_{n>0}(t^n) = (0)$. Thus, for any nonzero element x of A, there exists a unique nonnegative integer a with $x \in (t^a)$ and $x \notin (t^{a+1})$. If we write $x = ut^a$, then

---

[1] There are various equivalent definitions of a discrete valuation ring. See [18, Chapter 4] for details.
[2] Indeed, if we set $I = \bigcap_{n>0}(t^n)$, then I is a principal ideal by the assumption of A, so $I = (s)$ for some $s \in A$. Since $tI = I$, there exists $a \in A$ with $s = tas$. Then we obtain $s = 0$.

$u \notin (t)$, so $u \in A^{\times}$. In general, any element $x \neq 0$ in the fractional field of A is uniquely written as

$$x = t^a \cdot u, \quad (u \in A^{\times}, a \in \mathbb{Z}). \tag{2.10}$$

We denote a by $\mathrm{ord}_A(x)$, and call $\mathrm{ord}_A(\cdot)$ the *discrete valuation* of A.

Lemma 2.7 *Let K be a number field, let O_K be the ring of integers of K, and let P be a nonzero prime ideal of O_K. Then the localization $(O_K)_P$ of O_K at P is a discrete valuation ring.*

Proof We note that $(O_K)_P$ is a domain with unique maximal ideal $P(O_K)_P$, and that $(O_K)_P$ is not a field, because $P(O_K)_P \neq (0)$. Thus, it suffices to show that any ideal of $(O_K)_P$ is a principal ideal.

It follows from the existence of prime ideal decomposition in O_K (Theorem 2.5) that any ideal of $(O_K)_P$ is of the form $P^e(O_K)_P$, where e is a nonnegative integer. First, we claim that $P(O_K)_P$ is a principal ideal. To see this, we take $t \in P(O_K)_P$ with $t \notin P^2(O_K)_P$. Then the ideal (t) is of the form $P^e(O_K)_P$, and the choice of t implies $(t) = P(O_K)_P$. Next, we have $P^e(O_K)_P = (t^e)$ for any e. Hence, any ideal of $(O_K)_P$ is principal. $\qquad\square$

Finally, we show the following lemma, which we will use in Chapter 3. For a nonzero prime ideal P of O_K, we set $P \cap \mathbb{Z} = (p)$, where p is a prime of $\mathbb{Z}$. Since O_K is a free $\mathbb{Z}$-module of rank $[K : \mathbb{Q}]$, O_K/P is a finite extension of $\mathbb{Z}/(p)$ with degree at most $[K : \mathbb{Q}]$. In particular, O_K/P is a finite field, and $\#(O_K/P)$ is finite.

Lemma 2.8 *Let K be a number field, let O_K be the ring of integers of K, and let $I \neq (0)$ be an ideal of O_K. We write $I = P_1^{e_1} \cdots P_r^{e_r}$ for the prime ideal decomposition of I. Then*

$$\#(O_K/I) = \prod_{i=1}^{r} \#(O_K/P_i)^{e_i}.$$

Proof Let t_i be a generator of $P_i(O_K)_{P_i}$. Using the Chinese remainder theorem, we obtain

$$O_K/I \cong \bigoplus_{i=1}^{r} O_K/P_i^{e_i} = \bigoplus_{i=1}^{r} \left(O_K/P_i^{e_i}\right)_{P_i} = \bigoplus_{i=1}^{r} (O_K)_{P_i}/t_i^{e_i}(O_K)_{P_i}.$$

Since $(O_K)_{P_i}/t_i(O_K)_{P_i} \cong O_K/P_i$, and since the endomorphism of $(O_K)_{P_i}$-modules

$$(O_K)_{P_i}/t_i(O_K)_{P_i} \xrightarrow{\times t_i^k} t_i^k(O_K)_{P_i}/t_i^{k+1}(O_K)_{P_i}$$

is an isomorphism, we obtain

$$\#(O_K/I) = \prod_{i=1}^{r} \#\left((O_K)_{P_i}/t_i^{e_i}(O_K)_{P_i}\right)$$

$$= \prod_{i=1}^{r} \#\left((O_K)_{P_i}/t_i(O_K)_{P_i}\right)^{e_i} = \prod_{i=1}^{r} \#(O_K/P_i)^{e_i},$$

as required.   □

2.4 Lattices and Minkowski's Convex Body Theorem

In this section, we define a normed $\mathbb{R}$-vector space, and prove Minkowski's (Figure 2.2) convex body theorem (see Theorem 2.9). In the next section, we apply this theorem to a normed $\mathbb{R}$-vector space related to a number field and a lattice related to an ideal of the ring of integers (see Theorem 2.13).

Let V be an n-dimensional $\mathbb{R}$-vector space. A subset M of V is called a *lattice* if there are linearly independent vectors $e_1, \ldots, e_n \in V$ such that

$$M = \mathbb{Z}e_1 + \cdots + \mathbb{Z}e_n.$$

Then M is a free $\mathbb{Z}$-module of rank n and $\{e_1, \ldots, e_n\}$ is a free basis of M.

Assume that V is equipped with an inner product, i.e., a positive symmetric bilinear form

$$\langle \, , \, \rangle \colon V \times V \to \mathbb{R}.$$

Figure 2.2 Hermann Minkowski.
Source: https://en.wikipedia.org.

Given a free basis $\{e_1, \ldots, e_n\}$ of M, we consider the value

$$\sqrt{\det(\langle e_i, e_j \rangle)}.$$

Note that this value is independent of the choice of a free basis of M. Indeed, suppose that $\{e'_1, \ldots, e'_n\}$ is a free basis of M. We write $e'_j = \sum_{i=1}^{n} a_{ij} e_i$, and we set $A = (a_{ij})$. Then $A \in GL(n, \mathbb{Z})$ and we have

$$(\langle e'_i, e'_j \rangle) = {}^t A\, (\langle e_i, e_j \rangle)\, A.$$

Thus, we get $\sqrt{\det(\langle e_i, e_j \rangle)} = \sqrt{\det(\langle e'_i, e'_j \rangle)}$. We denote $\sqrt{\det(\langle e_i, e_j \rangle)}$ by

$$\mathrm{vol}(M, \langle\, , \,\rangle), \tag{2.11}$$

and call it the *volume* of M with respect to $\langle\, , \,\rangle$.

We explain why $\mathrm{vol}(M, \langle\, , \,\rangle)$ is called the volume of M. We take an orthonormal basis $\{u_1, \ldots, u_n\}$ of V. With respect to this basis, we identify V with the real vector space $\mathbb{R}^n$ and we equip V with the Lebesgue measure induced by the standard Lebesgue measure of $\mathbb{R}^n$. Then $\mathrm{vol}(M, \langle\, , \,\rangle)$ is equal to the volume of the n-dimensional parallelepiped Π spanned by $e_1, \ldots, e_n$, i.e.,

$$\Pi = \{s_1 e_1 + \cdots + s_n e_n \mid 0 \le s_1 < 1, \ldots, 0 \le s_n < 1\}. \tag{2.12}$$

Minkowski's convex body theorem, which we state below, is a very useful tool for studying properties of ideals of the ring of integers of an algebraic number field. A subset S of V is called a *convex body* if for any $x, y \in S$ and $0 \le t \le 1$, we have $tx + (1 - t)y \in S$. The subset S is said to be *centrally symmetric* if for any $x \in S$, we have $-x \in S$.

Theorem 2.9 (Minkowski's convex body theorem, Figure 2.3) *Let V be an n-dimensional normed $\mathbb{R}$-vector space, let M be a lattice of V, and let S be a*

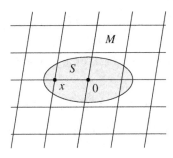

Figure 2.3 A centrally symmetric convex body S in a lattice M.

centrally symmetric convex body. Assume that $\mathrm{vol}(S) > 2^n \, \mathrm{vol}(M, \langle \, , \, \rangle)$. *Then there exists a nonzero element x of M such that* $x \in S$.

Proof We take a free basis $\{e_1, \ldots, e_n\}$ of M as a free $\mathbb{Z}$-module. We define Π by Equation (2.12).

It suffices to show that there exist distinct points $x_1, x_2 \in M$ such that

$$\left(\frac{1}{2}S - x_1\right) \cap \left(\frac{1}{2}S - x_2\right) \neq \emptyset.$$

Indeed, if $s_1, s_2 \in S$ satisfies $\frac{1}{2}s_1 - x_1 = \frac{1}{2}s_2 - x_2$, then we have $\frac{1}{2}s_1 + \frac{1}{2}(-s_2) = x_1 - x_2$. Since S is a centrally symmetric convex body, the left-hand side belongs to S and the right-hand side is a nonzero element of M.

To derive a contradiction, we assume that for any distinct points $x_1, x_2 \in M$, we have $\left(\frac{1}{2}S - x_1\right) \cap \left(\frac{1}{2}S - x_2\right) = \emptyset$. Then

$$\frac{1}{2^n} \mathrm{vol}(S) = \mathrm{vol}\left(\frac{1}{2}S\right) = \mathrm{vol}\left(\frac{1}{2}S \cap \coprod_{x \in M}(\Pi + x)\right)$$

$$= \sum_{x \in M} \mathrm{vol}\left(\frac{1}{2}S \cap (\Pi + x)\right) = \sum_{x \in M} \mathrm{vol}\left(\left(\frac{1}{2}S - x\right) \cap \Pi\right)$$

$$= \mathrm{vol}\left(\coprod_{x \in M}\left(\frac{1}{2}S - x\right) \cap \Pi\right) \leq \mathrm{vol}(\Pi) = \mathrm{vol}(M, \langle \, , \, \rangle),$$

which contradicts with the assumption. Hence, we obtain the assertion. $\qquad\square$

To end this section, we show the following lemma, which we will use later (see Lemma 2.12 and Theorem 3.3).

Lemma 2.10 *Let V be an n-dimensional normed $\mathbb{R}$-vector space and let M be a lattice in V. Let $x_1, \ldots, x_n$ be elements of M such that $\{x_1, \ldots, x_n\}$ is a basis of V. Then we have*

$$\sqrt{\det(\langle x_i, x_j \rangle)} = \#\left(\frac{M}{\mathbb{Z}x_1 + \cdots + \mathbb{Z}x_n}\right) \mathrm{vol}(M, \langle \, , \, \rangle).$$

Proof Let $\{e_1, \ldots, e_n\}$ be a free basis of M. We write $x_j = \sum_{i=1}^n b_{ij} e_i$, $B = (b_{ij}) \in M(n, n; \mathbb{Z})$. Then

$$(\langle x_i, x_j \rangle) = {}^t B \, (\langle e_i, e_j \rangle) \, B.$$

Since $\sqrt{\det(\langle e_i, e_j \rangle)} = \mathrm{vol}(M, \langle \, , \, \rangle)$, it suffices to show

$$\#\left(\frac{M}{\mathbb{Z}x_1 + \cdots + \mathbb{Z}x_n}\right) = |\det B|.$$

By the theory of elementary divisors, there are $P, Q \in \mathrm{GL}(n, \mathbb{Z})$ and $c_1, \ldots, c_n \in \mathbb{Z}$ such that

$$PBQ = \begin{pmatrix} c_1 & & 0 \\ & \ddots & \\ 0 & & c_n \end{pmatrix}.$$

We denote by f_P, f_B, f_Q the endomorphims of M as $\mathbb{Z}$-modules defined by P, B, Q, respectively. Then

$$\frac{M}{\mathbb{Z}x_1 + \cdots + \mathbb{Z}x_n} = \mathrm{Coker}(M \xrightarrow{f_B} M).$$

Since f_P and f_Q are isomorphisms of free $\mathbb{Z}$-modules, we have

$$\#\left(\frac{M}{\mathbb{Z}x_1 + \cdots + \mathbb{Z}x_n}\right) = \#\left(\mathrm{Coker}(M \xrightarrow{f_B} M)\right)$$

$$= \#\left(\mathrm{Coker}(M \xrightarrow{f_Q} M \xrightarrow{f_B} M \xrightarrow{f_P} M)\right)$$

$$= |c_1 \cdots c_n| = |\det B|.$$

This completes the proof. $\qquad\qquad\qquad\qquad\qquad\qquad\qquad\qquad\qquad\square$

2.5 Minkowski's Discriminant Theorem

Let K be a number field with $[K : \mathbb{Q}] = n$. For $x, y \in K$, we set

$$\langle x, y \rangle_K = \sum_{\sigma \in K(\mathbb{C})} \sigma(x)\bar{\sigma}(y),$$

where $\bar{\sigma}(y) = \overline{\sigma(y)}$. The complex conjugation gives an action on $K(\mathbb{C})$, so $\overline{\langle x, y \rangle_K} = \sum_{\sigma \in K(\mathbb{C})} \bar{\sigma}(x)\sigma(y) = \langle x, y \rangle_K$. Thus, $\langle x, y \rangle_K \in \mathbb{R}$. Similarly, since $\langle y, x \rangle_K = \langle x, y \rangle_K$, the form $\langle \, , \, \rangle_K$ is symmetric. It follows that if we regard K as a $\mathbb{Q}$-vector space, the form

$$\langle \, , \, \rangle_K : K \times K \to \mathbb{R}$$

is a symmetric bilinear form over $\mathbb{Q}$. Now we set

$$V = K \otimes_{\mathbb{Q}} \mathbb{R}. \qquad\qquad\qquad\qquad (2.13)$$

Since $[K : \mathbb{Q}] = n$, V is an n-dimensional $\mathbb{R}$-vector space. By linearity, we extend $\langle \, , \, \rangle_K$ to a symmetric bilinear form $V \times V \to \mathbb{R}$ over $\mathbb{R}$.

Proposition 2.11 *The form $\langle\,,\,\rangle_K$ is an inner product (i.e., a positive definite symmetric bilinear form) on V.*

Proof It is easy to see that $\langle\,,\,\rangle_K$ is symmetric and bilinear. We show that $\langle\,,\,\rangle_K$ is positive definite. We set $K(\mathbb{C}) = \{\sigma_1,\ldots,\sigma_n\}$. We fix a basis $\{\alpha_1,\ldots,\alpha_n\}$ of K over $\mathbb{Q}$, and we set $\Delta = (\sigma_i(\alpha_j))$.

For $x \in V$, we write

$$x = x_1\alpha_1 + \cdots + x_n\alpha_n \quad (x_1,\ldots,x_n \in \mathbb{R}).$$

Then

$$\langle x,x\rangle_K = \sum_{\sigma\in K(\mathbb{C})}\sum_{i=1}^{n}\sum_{j=1}^{n} x_i x_j \sigma(\alpha_i)\bar\sigma(\alpha_j) = {}^t(\Delta x)\overline{(\Delta x)} = |\Delta x|^2 \geq 0.$$

On the other hand, Equation (2.6) gives $D(\alpha_1,\ldots,\alpha_n) = \det(\Delta)^2$ and by Lemma 2.3 $D(\alpha_1,\ldots,\alpha_n) \neq 0$. Thus, $\det(\Delta) \neq 0$. It follows that if $\langle x,x\rangle_K = 0$, then $x = 0$.

We have shown that $\langle\,,\,\rangle_K$ is positive definite, so $\langle\,,\,\rangle_K$ gives an inner product of V. $\qquad\square$

Via the natural injective map

$$K \to V = K \otimes_{\mathbb{Q}} \mathbb{R}, \quad x \mapsto x \otimes 1,$$

we regard K as a subset of V.

Lemma 2.12 *Let I be a nonzero ideal of O_K. Then I is a lattice of V. Further, we have*

$$\mathrm{vol}(I,\langle\,,\,\rangle_K) = \#(O_K/I)\sqrt{|D_{K/\mathbb{Q}}|}.$$

In particular, $\mathrm{vol}(O_K,\langle\,,\,\rangle_K) = \sqrt{|D_{K/\mathbb{Q}}|}.$

Proof We set $K(\mathbb{C}) = \{\sigma_1,\cdots,\sigma_n\}$.

First we treat the case where $I = O_K$. Let $\{\omega_1,\ldots,\omega_n\}$ be an integral basis of O_K. Since $\{\omega_1,\ldots,\omega_n\}$ is a basis of K as a $\mathbb{Q}$-vector space, $\{\omega_1,\ldots,\omega_n\}$ is a basis of V as an $\mathbb{R}$-vector space. Thus, O_K is a lattice of V. We have $\langle\omega_i,\omega_j\rangle_K = \sum_{k=1}^{n}\sigma_k(\omega_i)\cdot\bar\sigma_k(\omega_j)$. If we set $\Delta = (\sigma_i(\omega_j))$, then we have

$$(\langle\omega_i,\omega_j\rangle_K) = {}^t\Delta\cdot\overline{\Delta}.$$

On the other hand, Equation (2.6) gives $D_{K/\mathbb{Q}} = (\det\Delta)^2$, so

$$\mathrm{vol}(O_K,\langle\,,\,\rangle_K) = \sqrt{\det(\langle\omega_i,\omega_j\rangle_K)} = |\det\Delta| = \sqrt{|D_{K/\mathbb{Q}}|}.$$

Next, we treat the general case. Let I be any nonzero ideal of O_K. By Lemma 2.8 and its preceding paragraph, $\#(O_K/I)$ is finite. Then I is a free $\mathbb{Z}$-module of rank n. We take a free basis $\{\alpha_1, \ldots, \alpha_n\}$ of I. Then $\{\alpha_1, \ldots, \alpha_n\}$ is a basis of V as an $\mathbb{R}$-vector space. Thus, I is a lattice of V. It follows from Lemma 2.10 that

$$\text{vol}(I, \langle\,,\,\rangle_K) = \#(O_K/I)\,\text{vol}(O_K, \langle\,,\,\rangle_K) = \#(O_K/I)\,\sqrt{|D_{K/\mathbb{Q}}|},$$

as required. $\square$

In the rest of this section, we apply Minkowski's convex body theorem to V so as to deduce Minkowski's discriminant theorem. We use this theorem to prove the Hermite–Minkowski theorem in Chapter 3 (see Theorem 3.38).

Theorem 2.13 (Minkowski's discriminant theorem) *Let K be a number field with $[K : \mathbb{Q}] = n$. We denote by $2r_2$ the number of embeddings of K to $\mathbb{C}$ that are not real. Then we have*

$$|D_{K/\mathbb{Q}}| \geq \left(\frac{\pi}{4}\right)^{2r_2} \frac{n^{2n}}{(n!)^2}.$$

Before we prove Theorem 2.13, we show several lemmas.
We denote by

$$\rho_1, \ldots, \rho_{r_1} : K \to \mathbb{R}$$

the elements of $K(\mathbb{C})$ such that the image of $K \to \mathbb{C}$ is in $\mathbb{R}$. We denote the rest by

$$\sigma_1, \bar{\sigma}_1, \ldots, \sigma_{r_2}, \bar{\sigma}_{r_2} : K \to \mathbb{C}.$$

Then

$$K(\mathbb{C}) = \{\rho_1, \ldots, \rho_{r_1}, \sigma_1, \bar{\sigma}_1, \ldots, \sigma_{r_2}, \bar{\sigma}_{r_2}\},$$

and $n = r_1 + 2r_2$.
We define a $\mathbb{Q}$-linear map $f : K \to \mathbb{R}^n$ as follows: For $x \in K$, we set

$$f(x) = \Big(\rho_1(x), \ldots, \rho_{r_1}(x), \sqrt{2}\,\text{Re}(\sigma_1(x)), \sqrt{2}\,\text{Im}(\sigma_1(x)), \ldots,$$
$$\sqrt{2}\,\text{Re}(\sigma_{r_2}(x)), \sqrt{2}\,\text{Im}(\sigma_{r_2}(x))\Big). \quad (2.14)$$

The map f induces an isomorphism of $\mathbb{R}$-vector spaces from $V = K \otimes_{\mathbb{Q}} \mathbb{R}$ to $\mathbb{R}^n$ [3]. For $x, y \in K$, we have

$$\langle x, y \rangle_K = \sum_{\sigma \in K(\mathbb{C})} \sigma(x)\bar{\sigma}(y) = \sum_{i=1}^{r_1} \rho_i(x)\rho_i(y) + \sum_{j=1}^{r_2} \sigma_j(x)\overline{\sigma_j(y)} + \sum_{j=1}^{r_2} \overline{\sigma_j(x)}\sigma_j(y).$$

The inner product $\langle\,,\,\rangle_K$ of V and the isomorphism $f : V \to \mathbb{R}^n$ induce an inner product $\langle\,,\,\rangle$ of $\mathbb{R}^n$ such that $\langle x, y \rangle_K = \langle f(x), f(y) \rangle$. Concretely, we have

$$\langle (a_1, \ldots, a_{r_1}, b_1, c_1, \ldots, b_{r_2}, c_{r_2}), (a'_1, \ldots, a'_{r_1}, b'_1, c'_1, \ldots, b'_{r_2}, c'_{r_2}) \rangle$$
$$= \sum_{i=1}^{r_1} a_i a'_i + \sum_{j=1}^{r_2} (b_j b'_j + c_j c'_j).$$

In other words, $\langle\,,\,\rangle$ is the standard inner product of $\mathbb{R}^n$. We endow $\mathbb{R}^n$ with the standard Lebesgue measure.

The next lemma is verified by straightforward computations.

Lemma 2.14 *Let t be a positive number. Let T_t be the subset of $\mathbb{R}^n$ defined by*

$$T_t = \left\{ (a_1, \ldots, a_{r_1}, b_1, c_1, \ldots, b_{r_2}, c_{r_2}) \in \mathbb{R}^n \,\middle|\, \sum_{i=1}^{r_1} |a_i| + \sum_{j=1}^{r_2} \sqrt{2(b_j^2 + c_j^2)} \leq t \right\}.$$

Then

$$\mathrm{vol}(T_t) = \frac{2^{r_1} \pi^{r_2}}{n!} t^n.$$

Proof   In general, for nonnegative integers k, l, we set

$$D(k,l)(t) = \left\{ (a_1, \ldots, a_k, s_1, \ldots, s_l) \in \mathbb{R}^{k+l} \,\middle|\, \begin{array}{l} a_i \geq 0, \ s_j \geq 0 \ (\forall i, j), \\ \sum_{i=1}^k a_i + \sum_{j=1}^l s_j \leq t \end{array} \right\},$$

and we set

$$I(k,l)(t) = \int_{D(k,l)(t)} s_1 \cdots s_l \, da_1 \cdots da_k ds_1 \cdots ds_l.$$

Then $I(k,l)(t) = t^{k+2l} I(k,l)(1)$.

---

[3] Indeed, for a basis of $\alpha_1, \ldots, \alpha_n$ of K as a $\mathbb{Q}$-vector space, it suffices to show that $f(\alpha_1), \ldots, f(\alpha_n)$ is $\mathbb{R}$-linearly independent. We assume that $\lambda_1 f(\alpha_1) + \cdots + \lambda_n f(\alpha_n) = 0$ $(\lambda_1, \ldots, \lambda_n \in \mathbb{R})$. It follows that, for any $\sigma \in K(\mathbb{C})$, we have $\lambda_1 \sigma(\alpha_1) + \cdots + \lambda_n \sigma(\alpha_n) = 0$. Since the matrix Δ in Proposition 2.11 is invertible, we obtain $\lambda_1 = \cdots = \lambda_n = 0$.

By the change of coordinates given by $b_j = \frac{1}{\sqrt{2}} s_j \cos(\theta_j)$, $c_j = \frac{1}{\sqrt{2}} s_j \sin(\theta_j)$ $(s_j \geq 0, \theta_j \in [0, 2\pi])$, we get

$$
\begin{aligned}
\operatorname{vol}(T_t) &= \int_{\substack{\sum_{i=1}^{r_1} |a_i| + \sum_{j=1}^{r_2} s_j \leq t \\ 0 \leq \theta_1 \leq 2\pi, \dots, 0 \leq \theta_{r_2} \leq 2\pi}} \frac{1}{2^{r_2}} s_1 \cdots s_{r_2} da_1 \cdots da_{r_1} ds_1 \cdots ds_{r_2} d\theta_1 \cdots d\theta_{r_2} \\
&= \pi^{r_2} \int_{\sum_{i=1}^{r_1} |a_i| + \sum_{j=1}^{r_2} s_j \leq t} s_1 \cdots s_{r_2} da_1 \cdots da_{r_1} ds_1 \cdots ds_{r_2} \\
&= 2^{r_1} \pi^{r_2} I(r_1, r_2)(t) = 2^{r_1} \pi^{r_2} t^n I(r_1, r_2)(1). \qquad (2.15)
\end{aligned}
$$

Now we compute $I(k, l)(1)$. Since

$$
\begin{aligned}
I(k, l)(1) &= \int_0^1 I(k-1, l)(1 - a_1) \, da_1 \\
&= \int_0^1 (1 - a_1)^{(k-1)+2l} I(k-1, l)(1) \, da_1 = \frac{1}{k + 2l} I(k-1, l)(1),
\end{aligned}
$$

we inductively obtain $I(k, l)(1) = \frac{(2l)!}{(k+2l)!} I(0, l)(1)$. Next, since

$$
\begin{aligned}
I(0, l)(1) &= \int_0^1 s_1 I(0, l-1)(1 - s_1) \, ds_1 \\
&= \int_0^1 s_1 (1 - s_1)^{2(l-1)} I(0, l-1)(1) \, ds_1 = \frac{1}{(2l)(2l-1)} I(0, l-1)(1),
\end{aligned}
$$

we inductively obtain $I(0, l)(1) = 1/(2l)!$. Thus, $I(k, l)(1) = 1/(k + 2l)!$, and Equation (2.15) gives the assertion. $\qquad \square$

Proof of Theorem 2.13 For a positive number t, we define the subset S_t of V by $S_t = f^{-1}(T_t)$. Then S_t is a centrally symmetric convex body. We have defined the inner product of $\mathbb{R}^n$ such that $f : V \to \mathbb{R}^n$ is an isometry. Thus,

$$
\operatorname{vol}(S_t) = \frac{2^{r_1} \pi^{r_2}}{n!} t^n.
$$

We recall from Lemma 2.12 that $\operatorname{vol}(O_K, \langle\ ,\ \rangle_K) = \sqrt{|D_{K/\mathbb{Q}}|}$. For any positive number ϵ, let t be a positive number that satisfies

$$
\frac{2^{r_1} \pi^{r_2}}{n!} t^n = 2^n \sqrt{|D_{K/\mathbb{Q}}|} + \epsilon.
$$

By Minkowski's convex body theorem (Theorem 2.9), there is a nonzero element $\alpha \in O_K$ with $\alpha \in S_t$.

It follows from the definition of S_t that "$\alpha \in S_t$" implies $\sum_{\sigma \in K(\mathbb{C})} |\sigma(\alpha)| < t$.
Further, it follows from $\alpha \in O_K$ that $N_{K/\mathbb{Q}}(\alpha) \in \mathbb{Z}$. Then the inequality of arithmetic and geometric means implies that

$$1 \leq |N_{K/\mathbb{Q}}(\alpha)| = \prod_{\sigma \in K(\mathbb{C})} |\sigma(\alpha)| \leq \left(\frac{1}{n} \sum_{\sigma \in K(\mathbb{C})} |\sigma(\alpha)| \right)^n$$
$$< \frac{t^n}{n^n} = \frac{1}{n^n} \frac{n!}{2^{r_1} \pi^{r_2}} \left(2^n \sqrt{|D_{K/\mathbb{Q}}|} + \epsilon \right).$$

Since ϵ is an arbitrary positive number, we get the assertion. $\qquad\square$

Remark 2.15   We put

$$f(n) = \left(\frac{\pi}{4} \right)^n \frac{n^{2n}}{(n!)^2}.$$

It follows from $2r_2 \leq n$ and $\pi/4 < 1$ that $|D_{K/\mathbb{Q}}| \geq f(n)$. Since

$$\frac{f(n+1)}{f(n)} = \left(\frac{\pi}{4} \right) \left(1 + \frac{1}{n} \right)^{2n} \geq \pi,$$

we get $f(n) \geq \pi^n/4$. In particular, if $K \neq \mathbb{Q}$, then $|D_{K/\mathbb{Q}}| > 1$. We also have

$$n \leq \frac{\log(4|D_{K/\mathbb{Q}}|)}{\log(\pi)},$$

which shows that the extension degree $n = [K : \mathbb{Q}]$ is bounded from above by a constant that depends only on the discriminant $|D_{K/\mathbb{Q}}|$.

2.6 Field Extension and Ramification Index

Let K be a number field. Let K' be a finite extension field of K. We write O_K and $O_{K'}$ for the rings of integers of K and K', respectively. Let P be a nonzero prime ideal of O_K, and let $P O_{K'} = P_1'^{e_1} \cdots P_r'^{e_r}$ be the prime decomposition in K'. The integer e_i is called the *ramification index* of K'/K at P_i'. The *residue degree* of K'/K at P_i' is defined by $f_i = \left[O_{K'}/P_i' : O_K/P \right]$.

If $e_i = 1$, then P_i' is said to be *unramified* over K. If $e_i \geq 2$, then P_i' is said to be *ramified* over K. We say that P is *unramified* if $P_1', \ldots, P_r'$ are all unramified, i.e., $e_1 = \cdots = e_r = 1$.

Lemma 2.16   *We have* $[K' : K] = e_1 f_1 + \cdots + e_r f_r$.

Proof Since $(O_K)_P$ is a principal ideal domain, $(O_{K'})_P$ is a free $(O_K)_P$-module of rank $[K' : K]$. Thus,

$$\dim_{O_K/P} O_{K'}/PO_{K'} = \dim_{O_K/P}(O_{K'})_P/P(O_{K'})_P \qquad (2.16)$$
$$= \dim_{O_K/P}((O_K)_P/P(O_K)_P) \otimes_{(O_K)_P} (O_{K'})_P$$
$$= [K' : K].$$

We put $a = \#(O_K/P)$. Then it follows from (2.16) and Lemma 2.8 that

$$[K' : K] = \log_a \#(O_{K'}/PO_{K'}) = \sum_{i=1}^{r} e_i \log_a \#(O_{K'}/P_i') = \sum_{i=1}^{r} e_i f_i,$$

as required. □

In the rest of this section, we explain how discriminants are related to ramifications of prime ideals.

As in Equation (2.4), we write $(\,,\,)_{\mathrm{Tr}_{K/\mathbb{Q}}} : K \times K \to \mathbb{Q}$ for the trace form and we define the fractional ideal $\mathcal{M}$ to be

$$\mathcal{M} = \{x \in K \mid (x,y)_{\mathrm{Tr}_{K/\mathbb{Q}}} \in \mathbb{Z} \text{ for any } y \in O_K\}.$$

We take an integral basis $\{\omega_1, \ldots, \omega_n\}$ of O_K, and let $\{\beta_1, \ldots, \beta_n\}$ be the dual basis with respect to $(\,,\,)_{\mathrm{Tr}_{K/\mathbb{Q}}}$. Then we have $\mathcal{M} = \mathbb{Z}\beta_1 + \cdots + \mathbb{Z}\beta_n$.

The *difference* of K is defined by $\mathcal{D}_K = \mathcal{M}^{-1}$. Since $O_K \subseteq \mathcal{M}$, we have $\mathcal{D}_K \subseteq O_K$, so $\mathcal{D}_K$ is an ideal of O_K. It is easy to see that[4] $\#(O_K/\mathcal{D}_K) = |D_{K/\mathbb{Q}}|$.

For $\alpha \in O_K$, let $f(X) \in \mathbb{Z}[X]$ be the minimal polynomial of α over $\mathbb{Q}$. Note that $f(X)$ is monic. We set

$$\delta_{K/\mathbb{Q}}(\alpha) = \begin{cases} f'(\alpha) & (\text{if } K = \mathbb{Q}(\alpha)), \\ 0 & (\text{if } K \neq \mathbb{Q}(\alpha)). \end{cases}$$

Lemma 2.17 *1. The ideal $\mathcal{D}_K$ is generated by $\{\delta_{K/\mathbb{Q}}(\alpha)\}_{\alpha \in O_K}$.*
2. For any nonzero prime ideal P of O_K, P is ramified over $\mathbb{Q}$ if and only if P appears in the prime decomposition of $\mathcal{D}_K$.

---

[4] Indeed, since $\#(O_K/\mathcal{D}_K) = \#(\mathcal{M}/O_K)$, it suffices to show $\#(\mathcal{M}/O_K) = |D_{K/\mathbb{Q}}|$. We write $\omega_j = \sum_{l=1}^{n} c_{lj}\beta_l$ ($c_{lj} \in \mathbb{Z}$). Then by the theory of elementary divisors, we have $\#(\mathcal{M}/O_K) = |\det(c_{lj})|$. On the other hand, we have $(\omega_i, \omega_j)_{\mathrm{Tr}_{K/\mathbb{Q}}} = \sum_{l=1}^{n} c_{lj}(\omega_i, \beta_l)_{\mathrm{Tr}_{K/\mathbb{Q}}} = c_{ij}$. Thus, $|\det(c_{ij})| = |\det((\omega_i, \omega_j)_{\mathrm{Tr}_{K/\mathbb{Q}}})| = |D_{K/\mathbb{Q}}|$. We conclude that $\#(O_K/\mathcal{D}_K) = |D_{K/\mathbb{Q}}|$.

3. *Let $p \in \mathbb{Z}$ be a prime number. We set $pO_K = P_1^{e_1} \cdots P_r^{e_r}$. Then we have* [5]

$$\mathrm{ord}_{P_i}(\mathcal{D}_K) \leq e_i - 1 + \mathrm{ord}_{P_i}(e_i).$$

For the proof of Lemma 2.17, we refer the reader to [23, Chapter III, Theorems (2.5), (2.6)].

Theorem 2.18 *Let K be a number field and let O_K be the ring of integers of K. We set $n = [K : \mathbb{Q}]$. Let $D_{K/\mathbb{Q}}$ be the discriminant of K over $\mathbb{Q}$. Suppose that S is a finite subset of primes of $\mathbb{Z}$ and that O_K is unramified outside of S over $\mathbb{Z}$. Then we have*

$$|D_{K/\mathbb{Q}}| \leq \prod_{p \in S} p^{n-1+n\log_p(n)}.$$

Proof Let $p \in S$ and we write $pO_K = P_1^{e_1} \cdots P_r^{e_r}$ for the prime decomposition of pO_K in O_K. Let $f_i = [O_K/P_i : \mathbb{Z}/p\mathbb{Z}]$ denote the residue degree of $K/\mathbb{Q}$ at P_i. Since $\mathrm{ord}_{P_i}(e_i) = e_i \, \mathrm{ord}_p(e_i)$, Lemma 2.17,3 gives

$$\log_p\left(\#\left((O_K)_p/(\mathcal{D}_K)_p\right)\right) = \sum_i \mathrm{ord}_{P_i}(\mathcal{D}_K) f_i$$

$$\leq \sum_i (e_i - 1 + e_i \, \mathrm{ord}_p(e_i)) f_i.$$

It follows from $p^{\mathrm{ord}_p(e_i)} \leq e_i \leq n$ that $\mathrm{ord}_p(e_i) \leq \log_p(n)$. Then using $\sum_i e_i f_i = n$ in Lemma 2.16, we obtain

$$\log_p\left(\#\left((O_K)_p/(\mathcal{D}_K)_p\right)\right) \leq \sum_i (e_i - 1 + e_i \log_p(n)) f_i$$

$$= n - r + n \log_p(n) \leq n - 1 + n \log_p(n).$$

Since $\#(O_K/\mathcal{D}_K) = \prod_{p \in S} \#\left((O_K)_p/(\mathcal{D}_K)_p\right)$, we obtain the assertion. $\qquad \square$

---

[5] For a fractional ideal I of K and a nonzero prime ideal P of O_K, we define $\mathrm{ord}_P(I)$ as follows: The localization of I at P, denoted by I_P, is given by $I_P = t_P^a(O_K)_P$ for some integer a. Here, t_P is a local parameter of $(O_K)_P$. We put $\mathrm{ord}_P(I) := a$. We define $\mathrm{ord}_{P_i}(e_i)$ as $\mathrm{ord}_{P_i}(e_i O_K)$ of the ideal $e_i O_K$ of O_K. With the notation in (2.10), $\mathrm{ord}_{P_i}(e_i) = \mathrm{ord}_{(O_K)_P}(e_i)$.

3

Theory of Heights

This chapter is devoted to the theory of heights, which is fundamental in Diophantine geometry. In particular, Northcott's finiteness theorem and the Mordell–Weil theorem are two of the most fundamental theorems in Diophantine geometry.

3.1 Absolute Values

We start with the notion of absolute values on a field.

Let F be a field. An *absolute value* (or a *multiplicative valuation*) on F is a map $|\cdot| : F \to \mathbb{R}$ satisfying

1. $|x| \geq 0$ for any $x \in F$, and $|x| = 0$ if and only if $x = 0$;
2. $|xy| = |x||y|$ for any $x, y \in F$;
3. (triangle inequality) $|x + y| \leq |x| + |y|$ for any $x, y \in F$.

An absolute value $|\cdot|$ is said to be *nonarchimedean* if in place of the triangle inequality 3, it satisfies a stronger triangle inequality

4. $|x + y| \leq \max\{|x|, |y|\}$ for any $x, y \in F$.

An absolute value that does not satisfy the inequality 4 is said to be *archimedean*.

Any field is equipped with the nonarchimedean absolute value defined by

$$|x| = \begin{cases} 1 & (\text{if } x \neq 0), \\ 0 & (\text{if } x = 0). \end{cases}$$

It is called the *trivial absolute value*.

Two absolute values $|\cdot|_1$ and $|\cdot|_2$ are said to be *equivalent* if there exists a positive number c such that $|\cdot|_1 = |\cdot|_2^c$.

We give a nonarchimedean absolute value on $\mathbb{Q}$. Let p be a prime. For a nonzero rational number x, we write

$$x = p^a \frac{n}{m} \quad (a, m, n \text{ are integers, and } p \text{ does not divide } m, n),$$

and we set $|x|_p = p^{-a}$. We also set $|0|_p = 0$. Then $|\cdot|_p$ is a nonarchimedean absolute value on $\mathbb{Q}$.

The usual absolute value on $\mathbb{Q}$ defined by

$$|x| = \begin{cases} x & (\text{if } x \geq 0), \\ -x & (\text{if } x < 0) \end{cases}$$

is an archimedean absolute value.

Let K be a number field. We construct a nonarchimedean absolute value on K as in the case of $\mathbb{Q}$. Let O_K be the ring of integers and let P be a nonzero prime ideal of O_K. By Lemma 2.7, the localization $(O_K)_P$ of O_K at P is a discrete valuation ring, so $P(O_K)_P = t_P(O_K)_P$ for some $t_P \in (O_K)_P$. For a nonzero element x of K, we write

$$x = t_P^a \cdot u \quad (a \text{ is an integer, and } u \in (O_K)_P^\times).$$

We denote a by $\mathrm{ord}_P(x)$. (See Equation (2.10), where we write $a = \mathrm{ord}_{(O_K)_P}(x)$.) We set

$$|x|_P = \#(O_K/P)^{-\mathrm{ord}_P(x)}.$$

Then $|\cdot|_P$ is a nonarchimedean absolute value on K. It is called the *normalized absolute value* at P.

We construct an archimedean absolute value on K as in the case of $\mathbb{Q}$. As in Equation (2.7), we denote the set of $\mathbb{C}$-valued points of K by

$$K(\mathbb{C}) = \{\sigma \mid \sigma : K \hookrightarrow \mathbb{C} \text{ is an embedding of fields}\}.$$

Then $\#(K(\mathbb{C})) = [K : \mathbb{Q}]$. For $\sigma \in K(\mathbb{C})$, we define $|x|_\sigma$ to be the usual absolute value of $\sigma(x)$ as a complex number, i.e.,

$$|x|_\sigma = \sqrt{\sigma(x)\overline{\sigma(x)}}.$$

Then $|\cdot|_\sigma$ is an archimedean absolute value on K.

Ostrowski's theorem asserts that these are the only nontrivial absolute values on K up to equivalence.

Theorem 3.1 (Ostrowski's theorem) *Let $|\cdot|$ be a nontrivial absolute value on a number field K. If $|\cdot|$ is nonarchimedean, then there exists a nonzero prime ideal P of O_K such that $|\cdot|$ is equivalent to $|\cdot|_P$. If $|\cdot|$ is archimedean, then there exists $\sigma \in K(\mathbb{C})$ such that $|\cdot|$ is equivalent to $|\cdot|_\sigma$.*

We do not use Theorem 3.1 in the sequel, so we omit a proof (see, for example, [23, Chapter II, Theorem (4.2)] for a proof).

Let K be a number field and let O_K be the ring of integers of K. We denote by $\mathrm{Spec}(O_K)$ the set of prime ideals of O_K. We set

$$M_K = (\mathrm{Spec}(O_K) \setminus \{(0)\}) \amalg K(\mathbb{C}).$$

For $v \in M_K$, we set

$$|\cdot|_v = \begin{cases} |\cdot|_P & (\text{if } v = P \in \mathrm{Spec}(O_K) \setminus \{(0)\}), \\ |\cdot|_\sigma & (\text{if } v = \sigma \in K(\mathbb{C})). \end{cases}$$

We call an element of M_K a *place*. A place v is said to be *archimedean* (resp. *nonarchimedean*) if the corresponding absolute value $|\cdot|_v$ is archimedean (resp. nonarchimedean). Archimedean (resp. nonarchimedean) places are also called *infinite* (resp. *finite*) places. We denote by M_K^∞ (resp. M_K^{fin}) the set of infinite (resp. finite) places.

Remark 3.2 For $\sigma \in K(\mathbb{C})$, we define $\bar\sigma$ by $\bar\sigma(x) = \overline{\sigma(x)}$ for any $x \in K$. Then $\bar\sigma \in K(\mathbb{C})$. If $\sigma(K) \not\subset \mathbb{R}$, then we have $\sigma \neq \bar\sigma$ as elements of $K(\mathbb{C})$, so σ and $\bar\sigma$ are different infinite places. However, the corresponding absolute values are the same: $|\cdot|_\sigma = |\cdot|_{\bar\sigma}$.

3.2 Product Formula

In this section, we prove the so-called *product formula*. It is an arithmetic analog of the formula that, for a meromorphic function on a compact Riemann surface, the degree of its principal divisor is zero. In Section 3.3, we use the product formula to define the height of an algebraic point on projective space. Let K be a number field.

Theorem 3.3 (product formula) $\prod_{v \in M_K} |x|_v = 1$ *for any $x \in K \setminus \{0\}$.*

Proof Noting that any $x \in K \setminus \{0\}$ is written as $x = x_1/x_2$ with $x_1, x_2 \in O_K \setminus \{0\}$, we may assume that $x \in O_K \setminus \{0\}$ to prove the product formula.

For $y, z \in K$, we set $\langle y, z \rangle_K = \sum_{\sigma \in K(\mathbb{C})} \sigma(y)\bar{\sigma}(z)$ as in Section 2.5. Then by Proposition 2.11, $\langle \, , \, \rangle_K$ extends to an inner product on the $\mathbb{R}$-vector space $V = K \otimes_{\mathbb{Q}} \mathbb{R}$. Further, by Lemma 2.12, O_K is a lattice of V.

We set $n = [K : \mathbb{Q}]$. Let $\{\omega_1, \ldots, \omega_n\}$ be an integral basis of O_K. Then $\{x\omega_1, \ldots, x\omega_n\}$ is a basis of V. It follows from Lemma 2.10 and $\mathbb{Z}x\omega_1 + \cdots + \mathbb{Z}x\omega_n = xO_K$ that

$$\sqrt{\det(\langle x\omega_i, x\omega_j \rangle_K)} = \# \left(\frac{O_K}{\mathbb{Z}x\omega_1 + \cdots + \mathbb{Z}x\omega_n} \right) \mathrm{vol}(O_K, \langle \, , \, \rangle_K)$$

$$= \#(O_K/xO_K)\, \mathrm{vol}(O_K, \langle \, , \, \rangle_K).$$

To finish the proof, it suffices to prove the following two equalities:

$$\det(\langle x\omega_i, x\omega_j \rangle_K) = \mathrm{vol}(O_K, \langle \, , \, \rangle_K)^2 \cdot \prod_{\sigma \in K(\mathbb{C})} |x|_\sigma^2, \tag{3.1}$$

$$\#(O_K/xO_K) \cdot \prod_{P \in \mathrm{Spec}(O_K) \setminus \{(0)\}} |x|_P = 1. \tag{3.2}$$

First, we show Equation (3.1). We write $K(\mathbb{C}) = \{\sigma_1, \ldots, \sigma_n\}$. Then

$$\langle x\omega_i, x\omega_j \rangle_K = \sum_{k=1}^n \sigma_k(x\omega_i) \cdot \bar{\sigma}_k(x\omega_j) = \sum_{k=1}^n |x|_{\sigma_k}^2 \sigma_k(\omega_i) \cdot \bar{\sigma}_k(\omega_j).$$

Putting $\Delta = (\sigma_i(\omega_j))$, we obtain

$$(\langle x\omega_i, x\omega_j \rangle_K) = {}^t\Delta \begin{pmatrix} |x|_{\sigma_1}^2 & & 0 \\ & \ddots & \\ 0 & & |x|_{\sigma_n}^2 \end{pmatrix} \overline{\Delta}.$$

On the other hand, by Lemma 2.12, $\mathrm{vol}(O_K, \langle \, , \, \rangle_K)^2 = |\det(\Delta)|^2$, so we obtain Equation (3.1).

Equation (3.2) is a corollary of Lemma 2.8. Indeed, let $xO_K = P_1^{e_1} \cdots P_r^{e_r}$ be the prime ideal decomposition of xO_K. Then

$$\mathrm{ord}_P(x) = \begin{cases} e_i & (\text{if } P = P_i), \\ 0 & (\text{if } P \notin \{P_1, \ldots, P_r\}). \end{cases}$$

By Lemma 2.8, we have

$$\#(O_K/xO_K) = \prod_{i=1}^r \#(O_K/P_i)^{e_i} = \prod_{i=1}^r |x|_{P_i}^{-1} = \prod_{P \in \mathrm{Spec}(O_K) \setminus \{(0)\}} |x|_P^{-1}.$$

We have shown the product formula. $\qquad\square$

Remark 3.4   Taking the logarithm in the product formula (Theorem 3.3), we obtain

$$\sum_{v \in M_K} \log(|x|_v) = 0.$$

The product formula is often used in this form. Also, we see from this form that the product formula is an analogy of the fact that the degree of the principal divisor of a meromorphic function on a compact Riemann surface is equal to 0.

3.3 Heights of Vectors and Points in Projective Space

In this section, we define the heights of vectors and the heights of points in projective space. Let K be a number field and let $x = (x_1, \ldots, x_n) \in K^n$. If $x \neq 0$, we define the *absolute (logarithmic) Weil height* of x (also simply called the *height* of x) by

$$h_K(x) = \frac{1}{[K : \mathbb{Q}]} \sum_{v \in M_K} \log \left(\max_{1 \leq i \leq n} \{|x_i|_v\} \right).$$

Further, we define

$$h_K^+(x) = \frac{1}{[K : \mathbb{Q}]} \sum_{v \in M_K} \log^+ \left(\max_{1 \leq i \leq n} \{|x_i|_v\} \right).$$

Here, $\log^+ : [0, \infty) \to \mathbb{R}$ is a continuous function defined by

$$\log^+(a) = \begin{cases} 0 & (\text{if } a < 1), \\ \log(a) & (\text{if } a \geq 1). \end{cases}$$

We remark that the function $\log^+$ satisfies

$$\log^+(ab) \leq \log^+(a) + \log^+(b)$$

for all $a, b \geq 0$. We also remark that h_K^+ is defined even for the zero vector as $h_K^+(0, \ldots, 0) = 0$. The functions h_K and h_K^+ are related as

$$h_K^+(x_1, \ldots, x_n) = h_K(1, x_1, \ldots, x_n).$$

Geometrically, this equality connects heights of points in affine space with those in projective space.

Let us look at the basic properties of heights.

Proposition 3.5   *1. Let K' be a finite extension of K. Let x be a nonzero vector of K^n. Then the height of x as a vector of K^n is the same as that as a vector of K'^n. In other words, $h_K(x) = h_{K'}(x)$.*

2. *h_K is invariant under scalar multiplication, i.e., for any nonzero element a of K and for any nonzero vector $x \in K^n$, we have $h_K(ax) = h_K(x)$.*
3. *For any nonzero vector $x \in K^n$, we have $h_K(x) \geq 0$.*
4. *For any $a \in K \setminus \{0\}$ and $n \in \mathbb{Z}$, we have $h_K^+(a^n) = |n| h_K^+(a)$.*
5. *For any $x = (x_1, \ldots, x_n) \in O_K^n \setminus \{0\}$, we have*

$$h_K(x) \leq \sum_{\sigma \in K(\mathbb{C})} \log \left(\max_{1 \leq i \leq n} \{|x_i|_\sigma\} \right).$$

Before we give a proof of Proposition 3.5, we explain some consequences. By Proposition 3.5, 1, for $x \in \overline{\mathbb{Q}}^n \setminus \{0\}$, if we take a number field K with $x \in K^n$, then $h_K(x)$ is independent of the choice of K. In the following, we will omit the suffix K, and simply write $h(x)$ for the height of x. Thus, we have a well-defined height function

$$h \colon \overline{\mathbb{Q}}^n \setminus \{0\} \to \mathbb{R}.$$

Further, by 2, for $x \in \mathbb{P}^{n-1}(\overline{\mathbb{Q}})$, the value $h(x)$ is well-defined, and we have a height function on projective space $\mathbb{P}^{n-1}$

$$h \colon \mathbb{P}^{n-1}(\overline{\mathbb{Q}}) \to \mathbb{R}. \tag{3.3}$$

As in the beginning of this section, the height h on projective space in (3.3) is also called the absolute (logarithmic) Weil height.

As for h_K^+, since $h_K^+(x)$ is also independent of the choice of K, we will similarly omit the suffix K in the following, and simply write $h^+(x)$.

Proof of Proposition 3.5   1. Let P be a prime ideal of O_K, and let $P O_{K'} = P_1'^{e_1} \cdots P_r'^{e_r}$ be the prime decomposition of $P O_{K'}$. For each x_j, we set $a_j = \operatorname{ord}_P(x_j)$. Then $\operatorname{ord}_{P_i'}(x_j) = a_j e_i$. Let $f_i = [O_{K'}/P_i' : O_K/P]$ be the residue degree. Then

$$|x_j|_{P_1'} \cdots |x_j|_{P_r'} = \#(O_{K'}/P_1')^{-\operatorname{ord}_{P_1'}(x_j)} \cdots \#(O_{K'}/P_r')^{-\operatorname{ord}_{P_r'}(x_j)}$$
$$= \#(O_K/P)^{-f_1 \operatorname{ord}_{P_1'}(x_j) - \cdots - f_r \operatorname{ord}_{P_r'}(x_j)}$$
$$= \#(O_K/P)^{-(e_1 f_1 + \cdots + e_r f_r) a_j},$$

and it follows from Lemma 2.16 that

$$|x_j|_{P_1'} \cdots |x_j|_{P_r'} = |x_j|_P^{[K':K]}. \tag{3.4}$$

On the other hand, for $\sigma \in K(\mathbb{C})$, we put

$$K'(\mathbb{C})_\sigma = \{\sigma' \in K'(\mathbb{C}) \mid \sigma'\big|_K = \sigma\}.$$

Then $\#(K'(\mathbb{C})_\sigma) = [K' : K]$, and we have

$$\prod_{\sigma' \in K'(\mathbb{C})_\sigma} |x_j|_{\sigma'} = |x_j|_\sigma^{[K':K]}. \tag{3.5}$$

Assertion 1 is the consequence of Equations (3.4) and (3.5).

2. By the definition of heights and the product formula (Theorem 3.3), we have

$$h_K(ax) = \frac{1}{[K : \mathbb{Q}]} \sum_v \log(|a|_v) + h_K(x) = h_K(x).$$

3. We put $x = (x_1, \ldots, x_n)$. Since x is a nonzero vector, $x_i \neq 0$ for some i. We put $y_j = x_j/x_i$ $(j = 1, \ldots, n)$ and $y = (y_1, \ldots, y_n)$. Then $y_i = 1$ and $x = x_i y$. By 2, we have $h_K(x) = h_K(y)$. Since $y_i = 1$, we have $\max_{1 \le j \le n}\{|y_j|_v\} \ge 1$. Thus, we obtain the assertion.

4. Suppose that $n \ge 0$. Since $\log^+(\alpha^n) = n\log^+(\alpha)$ for any positive number $\alpha > 0$, the assertion is obvious. Suppose now that $n < 0$. Then by 2 and the case of $n \ge 0$, we get

$$h_K^+(a^n) = h_K(1, a^n) = h_K(a^n(a^{-n}, 1)) = h_K(a^{-n}, 1)$$
$$= h_K^+(a^{-n}) = (-n)h_K^+(a) = |n|h_K^+(a).$$

5. If $a \in O_K$ and v is nonarchimedean, then $|a|_v \le 1$. Then the assertion follows from the definition of h_K. □

Let us show some more properties of heights.

Proposition 3.6   *1. For vectors $x = (x_i) \in \overline{\mathbb{Q}}^n \setminus \{0\}$ and $y = (y_j) \in \overline{\mathbb{Q}}^m \setminus \{0\}$, let $x \otimes y \in \overline{\mathbb{Q}}^{nm} \setminus \{0\}$ be the vector with coordinates $x_i y_j$, i.e., $x \otimes y = (x_i y_j)_{1 \le i \le n, 1 \le j \le m}$. Then $h(x \otimes y) = h(x) + h(y)$. In particular, $h(x^{\otimes m}) = mh(x)$.*
2. *Let $\phi: \overline{\mathbb{Q}}^n \to \overline{\mathbb{Q}}^m$ be a linear map over $\overline{\mathbb{Q}}$. Then there exists a constant C that depends only on ϕ such that for any $x \in \overline{\mathbb{Q}}^n \setminus \{0\}$ with $\phi(x) \neq 0$, we have*

$$h(\phi(x)) \le h(x) + C.$$

3. *Let $\phi: \overline{\mathbb{Q}}^n \to \overline{\mathbb{Q}}^n$ be an invertible linear map over $\overline{\mathbb{Q}}$. Then there is a constant C that depends only on ϕ such that, for any $x \in \overline{\mathbb{Q}}^n \setminus \{0\}$, we have*

$$|h(x) - h(\phi(x))| \le C.$$

Proof 1. We take a number field K with $x_i, y_j \in K$. For each $v \in M_K$, we have

$$\max_{i,j}\{|x_i y_j|_v\} = \max_i\{|x_i|_v\} \max_j\{|y_j|_v\}.$$

The assertion follows.

2. Let $(a_{ij}) \in M(m, n; \overline{\mathbb{Q}})$ be the matrix representation of ϕ with respect to the standard bases of $\overline{\mathbb{Q}}^n$ and $\overline{\mathbb{Q}}^m$. Since $\phi(x) \neq 0$, we have $(a_{ij}) \neq 0$. We put $C = h((a_{ij})) + \log(n)$. We claim that

$$h(\phi(x)) \leq h(x) + C$$

for any $x \in \overline{\mathbb{Q}}^n$. To see this, we take a number field K with $a_{ij} \in K$, $x \in K^n$. We write $x = {}^t(x_1, \ldots, x_n)$, then the i-th coordinate of $\phi(x)$ is given by $\sum_k a_{ik} x_k$. Here we have

$$\left| \sum_k a_{ik} x_k \right|_v \leq \begin{cases} \max_{i,j}\{|a_{ij}|_v\} \max_i\{|x_i|_v\} & (\text{if } v \in M_K^{\text{fin}}), \\ n \max_{i,j}\{|a_{ij}|_v\} \max_i\{|x_i|_v\} & (\text{if } v \in M_K^{\infty}). \end{cases}$$

Thus, we get $h(\phi(x)) \leq h(x) + C$.

3. Let $(b_{ij}) \in \mathrm{GL}(n, \overline{\mathbb{Q}})$ be the matrix representation of the inverse ϕ^{-1} of ϕ with respect to the standard basis of $\overline{\mathbb{Q}}^n$. We set $C' = h((b_{ij})) + \log(n)$. Then it follows from 2 that

$$h(\phi^{-1}(x)) \leq h(x) + C'$$

for any $x \in \overline{\mathbb{Q}}^n$. Now replacing x by $\phi(x)$, we have

$$h(x) \leq h(\phi(x)) + C'.$$

Then replacing C in 2 with $\max\{C, C'\}$, we obtain $|h(x) - h(\phi(x))| \leq C$. □

Proposition 3.6, 3 carries an important implication. Let V be a vector space of dimension $n + 1$ over $\overline{\mathbb{Q}}$. Let $x \in \mathbb{P}(V)$. If we fix a basis of V and identify $\mathbb{P}(V)$ with $\mathbb{P}^n$, then the absolute (logarithmic) Weil height $h(x)$ of x is defined. However, the value $h(x)$ depends on the choice of a basis of V. In other words, the absolute (logarithmic) Weil height is not invariant under linear coordinate changes. This is why it is sometimes called the *naive* height.

The naive height is not invariant under linear coordinate changes, but Proposition 3.6, 3 shows that a linear coordinate change alters the naive height with only up to a bounded function.

To give a sophisticated definition of a height is an important theme in arithmetic geometry. In later sections, we show that on abelian varieties we can define a better height, called the *Néron–Tate height*, without ambiguity of bounded functions.

3.4 Height Functions Associated to Line Bundles

In this section, we will generalize the absolute (logarithmic) Weil height functions on projective spaces in the previous section, and we will define height functions associated to line bundles on projective varieties. As we remarked in the previous section, the absolute (logarithmic) Weil functions are not invariant under linear coordinate changes. As a consequence, height functions associated to line bundles will be defined only up to modulo bounded functions.

First, we recall a projection from a linear subspace of $\mathbb{P}^n$. We give an explicit description. Let m be an integer with $0 \le m < n$, and we fix a linear subspace Δ of $\mathbb{P}^n$ with dimension $n - m - 1$ over $\overline{\mathbb{Q}}$. Let $\pi : \overline{\mathbb{Q}}^{n+1} \setminus \{0\} \to \mathbb{P}^n(\overline{\mathbb{Q}})$ be the natural quotient map. We take a basis $\{\omega_0, \ldots, \omega_n\}$ of $\overline{\mathbb{Q}}^{n+1}$ such that $\{\omega_{m+1}, \ldots, \omega_n\}$ gives a basis of the linear subspace $\pi^{-1}(\Delta) \cup \{0\}$ of $\overline{\mathbb{Q}}^{n+1}$. Let $\{\omega_0^\vee, \ldots, \omega_n^\vee\}$ be the dual basis of $\{\omega_0, \ldots, \omega_n\}$. Note that

$$\Delta = \{\pi(x) \mid x \in \overline{\mathbb{Q}}^{n+1} \setminus \{0\}, \ \omega_0^\vee(x) = \cdots = \omega_m^\vee(x) = 0\}.$$

Then the projection centered at Δ (Figure 3.1), denoted by

$$\phi_\Delta : \mathbb{P}^n(\overline{\mathbb{Q}}) \setminus \Delta \to \mathbb{P}^m(\overline{\mathbb{Q}}), \tag{3.6}$$

is defined to be

$$\phi_\Delta(\pi(x)) = (\omega_0^\vee(x) : \cdots : \omega_m^\vee(x)) \tag{3.7}$$

for any $x \in \overline{\mathbb{Q}}^{n+1} \setminus (\pi^{-1}(\Delta) \cup \{0\})$.

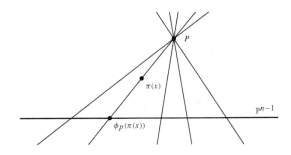

Figure 3.1 The case where Δ is a point p.

Note that ϕ_Δ in (3.6) depends on the choice of a basis $\{\omega_0, \ldots, \omega_n\}$ of $\overline{\mathbb{Q}}^{n+1}$. The definition of the projection ϕ_Δ as above is an extrinsic one. There is also an intrinsic definition of the projection, which we briefly explain. We denote by $\mathbb{P}^n(\overline{\mathbb{Q}})_\Delta$ the set of all linear subspaces of $\mathbb{P}^n(\overline{\mathbb{Q}})$ of dimension $n - m$ that contain Δ. Then we define

$$\phi_\Delta : \mathbb{P}^n(\overline{\mathbb{Q}}) \setminus \Delta \to \mathbb{P}^n(\overline{\mathbb{Q}})_\Delta \qquad (3.8)$$

by, to each $x \in \mathbb{P}^n(\overline{\mathbb{Q}}) \setminus \Delta$, associating an element of $\mathbb{P}^n(\overline{\mathbb{Q}})_\Delta$ that passes at x. For the map ϕ_Δ in (3.8), we do not need to choose a basis of $\overline{\mathbb{Q}}^{n+1}$. However, to show that $\mathbb{P}^n(\overline{\mathbb{Q}})_\Delta$ is isomorphic to $\mathbb{P}^m(\overline{\mathbb{Q}})$, we need to choose a basis of $\overline{\mathbb{Q}}^{n+1}$.

Lemma 3.7 *Let X be a projective subvariety of $\mathbb{P}^n$ over $\overline{\mathbb{Q}}$. Let Δ be a linear subspace of $\mathbb{P}^n(\overline{\mathbb{Q}})$ that is disjoint from $X(\overline{\mathbb{Q}})$. Choosing a basis of $\overline{\mathbb{Q}}^{n+1}$, let $\phi_\Delta : \mathbb{P}^n(\overline{\mathbb{Q}}) \setminus \Delta \to \mathbb{P}^m(\overline{\mathbb{Q}})$ be the projection from Δ and let $\phi : X(\overline{\mathbb{Q}}) \to \mathbb{P}^m(\overline{\mathbb{Q}})$ be the restriction to $X(\overline{\mathbb{Q}})$. Then there is a constant C such that, for any $x \in X(\overline{\mathbb{Q}})$,*

$$|h(x) - h(\phi(x))| \le C.$$

Proof By Proposition 3.6, 3, we may assume that the basis $\{\omega_0, \ldots, \omega_n\}$ of $\overline{\mathbb{Q}}^{n+1}$ that we have chosen to define ϕ_Δ is the standard basis. Then ϕ_Δ is given by $(x_0 : \cdots : x_n) \mapsto (x_0 : \ldots : x_m)$. We define $\psi_t : \mathbb{P}^t(\overline{\mathbb{Q}}) \dashrightarrow \mathbb{P}^{t-1}(\overline{\mathbb{Q}})$ by $\psi_t(x_0 : \cdots : x_t) = (x_0 : \cdots : x_{t-1})$. Since $\phi_\Delta = \psi_{m+1} \circ \cdots \circ \psi_n$, it suffices to show the case where ϕ_Δ is the projection centered at $P = (0 : \cdots : 0 : 1)$, i.e., $\Delta = P$ and

$$\phi_P : \mathbb{P}^n(\overline{\mathbb{Q}}) \setminus \{P\} \to \mathbb{P}^{n-1}(\overline{\mathbb{Q}}), \quad (x_0 : \cdots : x_n) \mapsto (x_0 : \cdots : x_{n-1}).$$

For a sequence of nonnegative integers $I = (i_0, \ldots, i_n)$, we write $\boldsymbol{X}^I = X_0^{i_0} \cdots X_n^{i_n}$ and $|I| = i_0 + \cdots + i_n$. Since $P \notin X$, there is a homogeneous polynomial F such that F vanishes on X and $F(P) \ne 0$. Let d be the degree of F. Replacing F by a constant multiple of F if necessary, we may assume that

$$F = X_n^d - \sum_{\substack{|I|=d, \\ I \ne (0,\ldots,0,d)}} a_I \boldsymbol{X}^I \quad (a_I \in \overline{\mathbb{Q}}).$$

We set $C = h^+((a_I)) + \log\left(\binom{n+d}{n} - 1\right)$. Let x be any point in $X(\overline{\mathbb{Q}})$. Since $X(\overline{\mathbb{Q}}) \subset \mathbb{P}^n(\overline{\mathbb{Q}})$, we write $x = (x_0 : \cdots : x_n)$. We set $x^I = x_0^{i_0} \cdots x_n^{i_n}$. We claim that

$$h\left((x^I)_{|I|=d}\right) \le h\left((x^I)_{\substack{|I|=d, \\ I \ne (0,\ldots,0,d)}}\right) + C. \qquad (3.9)$$

To see this, we take a number field K such that $x \in X(K)$ and $a_I \in K$ for any I. Then, since $x_n^d = \sum_{I \neq (0,\dots,0,d)} a_I x^I$, we have

$$|x_n^d|_v \leq \begin{cases} \max_{I \neq (0,\dots,0,d)}\{|a_I|_v\} \\ \qquad \times \max_{I \neq (0,\dots,0,d)}\{|x^I|_v\}, & \text{(if } v \in M_K^{\mathrm{fin}}), \\[2ex] \left(\binom{n+d}{n} - 1\right) \max_{I \neq (0,\dots,0,d)}\{|a_I|_v\} \\ \qquad \times \max_{I \neq (0,\dots,0,d)}\{|x^I|_v\}, & \text{(if } v \in M_K^{\infty}), \end{cases}$$

from which (3.9) follows.

On the other hand, we have

$$h\left((x^I)_{|I|=d}\right) = h(x^{\otimes d}).$$

As $P \notin X$, we have $x' = (x_0 : \cdots : x_{n-1}) \in \mathbb{P}^{n-1}(\overline{\mathbb{Q}})$, and

$$h\left((x^I)_{\substack{|I|=d, \\ I \neq (0,\dots,0,d)}}\right) = h(x^{\otimes d-1} \otimes x').$$

It follows from (3.9) and Proposition 3.6,1 that $dh(x) \leq (d-1)h(x) + h(x') + C$, i.e., $h(x) \leq h(x') + C$. Since $h(x') \leq h(x)$ is obvious, we obtain the assertion. $\qquad \square$

Let X be a projective variety over $\overline{\mathbb{Q}}$, and let $\phi : X \to \mathbb{P}^m$ be a morphism over $\overline{\mathbb{Q}}$. We define

$$h_\phi : X(\overline{\mathbb{Q}}) \to \mathbb{R}$$

by $h_\phi(x) = h(\phi(x))$ for $x \in X(\overline{\mathbb{Q}})$. We call h_ϕ the *height associated to* ϕ. Then we have the following:

Proposition 3.8 *We consider two morphisms* $\phi_1 : X \to \mathbb{P}^{m_1}$ *and* $\phi_2 : X \to \mathbb{P}^{m_2}$ *over* $\overline{\mathbb{Q}}$. *If* $\phi_1^*(\mathcal{O}_{\mathbb{P}^{m_1}}(1)) \cong \phi_2^*(\mathcal{O}_{\mathbb{P}^{m_2}}(1))$, *then there is a constant* C *such that, for any* $x \in X(\overline{\mathbb{Q}})$,

$$|h_{\phi_1}(x) - h_{\phi_2}(x)| \leq C.$$

Proof We set $L = \phi_1^*(\mathcal{O}_{\mathbb{P}^{m_1}}(1))$. Let $\{t_0, \dots, t_m\}$ be a basis of $H^0(X, L)$, and we set $\phi = (t_0 : \cdots : t_m)$. It suffices to show that there is a constant C_1 such that, for any $x \in X(\overline{\mathbb{Q}})$,

$$|h_\phi(x) - h_{\phi_1}(x)| \leq C_1. \tag{3.10}$$

Indeed, if this is the case, we similarly see that there is a constant C_2 such that, for any $x \in X(\overline{\mathbb{Q}})$, $|h_\phi(x) - h_{\phi_2}(x)| \leq C_2$. Then the triangular inequality implies the assertion with $C = C_1 + C_2$.

Let $X_0, \ldots, X_{m_1}$ be the homogeneous coordinates of $\mathbb{P}^{m_1}$, and we set $s_i = \phi_1^*(X_i) \in H^0(X, L)$. Rearranging $s_0, \ldots, s_{m_1}$ if necessary, we may assume that $s_0, \ldots, s_r$ are linearly independent and that each of $s_{r+1}, \ldots, s_{m_1}$ is a linear combinations of $s_0, \ldots, s_r$. We set

$$\phi_1' = (s_0 : \cdots : s_r) : X \to \mathbb{P}^r.$$

Then $h(\phi_1'(x)) \le h(\phi_1(x))$. On the other hand, Proposition 3.6, 2 tells us that there is a constant C' such that $h(\phi_1(x)) \le h(\phi_1'(x)) + C'$. Thus, for any $x \in X(\overline{\mathbb{Q}})$, $|h_{\phi_1'}(x) - h_{\phi_1}(x)| \le C'$.

We take $s_{r+1}', \ldots, s_m' \in H^0(X, L)$ such that $\{s_0, \ldots, s_r, s_{r+1}', \ldots, s_m'\}$ gives a basis for $H^0(X, L)$. We set $\phi' = (s_0 : \cdots : s_r : s_{r+1}' : \cdots : s_m') : X \to \mathbb{P}^m$. Then ϕ_1' is the composition of ϕ' and the projection $(X_0 : \cdots : X_m) \mapsto (X_0 : \cdots : X_{m_1})$. We apply Lemma 3.7 to $\phi'(X)$. Then there exists a constant C'' such that, for any $x \in X(\overline{\mathbb{Q}})$, $|h_{\phi_1'}(x) - h_{\phi'}(x)| \le C''$.

Finally, since $\{s_0, \ldots, s_r, s_{r+1}', \ldots, s_m'\}$ and $\{t_0, \ldots, t_m\}$ are both bases of $H^0(X, L)$, it follows from Proposition 3.6, 3 that there is a constant C''' such that, for any $x \in X(\overline{\mathbb{Q}})$, $|h_{\phi'}(x) - h_{\phi}(x)| \le C'''$. Then (3.10) holds with $C_1 = C' + C'' + C'''$. We obtain the assertion. $\qquad\square$

Let X be a projective variety over $\overline{\mathbb{Q}}$. Let $\mathrm{Func}(X)$ denote the set of all real-valued functions on $X(\overline{\mathbb{Q}})$, and let $B(X)$ denote the set of all bounded real-valued functions on $X(\overline{\mathbb{Q}})$. In the following, for two functions $h_1, h_2 \in \mathrm{Func}(X)$, we write

$$h_1 = h_2 + O(1)$$

if $h_1 - h_2 \in B(X)$. In other words, $h_1 = h_2 + O(1)$ if and only if there is a constant C that may depend on h_1, h_2 such that, for any $x \in X(\overline{\mathbb{Q}})$, $|h_1(x) - h_2(x)| \le C$.

Theorem 3.9 *To any projective variety X over $\overline{\mathbb{Q}}$ and to any line bundle L on X, we can uniquely attach $h_L \in \mathrm{Func}(X)$ modulo $B(X)$ with the following properties:*

1. *For any line bundles L_1, L_2 on X, we have $h_{L_1 \otimes L_2} = h_{L_1} + h_{L_2} + O(1)$.*
2. *For any morphism $f : X \to Y$ of projective varieties and for any line bundle L on Y, we have $h_{f^*(L)} = h_L \circ f + O(1)$.*
3. *For any morphism $\phi : X \to \mathbb{P}^n$, we have $h_{\phi^*(\mathcal{O}_{\mathbb{P}^n}(1))} = h_\phi + O(1)$.*

We call h_L a *height function associated to a line bundle L.*

Proof   We divide the proof into five steps.

Step 1. First, suppose that L is globally generated. Let

$$\phi_{|L|} : X \to \mathbb{P}(H^0(X, L))$$

be a morphism associated to the complete linear system $|L|$. We set $h_L = h_{\phi_{|L|}}$. Then, by Proposition 3.8, h_L satisfies property 3.

Step 2. For a globally generated line bundle L, we have defined h_L as in Step 1. Let us verify properties 1 and 2 for globally generated line bundles.

To verify property 1, suppose that L_1, L_2 are globally generated, let $\{s_i\}$ be a basis of $H^0(X, L_1)$, and let $\{t_j\}$ be a basis of $H^0(X, L_2)$. Then $\{s_i \otimes t_j\}$ induces a morphism $\phi : X \to \mathbb{P}^N$ such that $\phi^*(\mathcal{O}_{\mathbb{P}^N}(1)) = L_1 \otimes L_2$. It follows from Proposition 3.6, 1 that $h_\phi = h_{L_1} + h_{L_2} + O(1)$. On the other hand, Proposition 3.8 gives $h_\phi = h_{L_1 \otimes L_2} + O(1)$. Thus, we obtain property 1.

To verify property 2, suppose that L is globally generated, and let $\phi' : Y \to \mathbb{P}^m$ be a morphism associated to $|L|$. Since $f^*(\phi'^*(\mathcal{O}_{\mathbb{P}^m}(1))) = f^*(L)$, Proposition 3.8 gives

$$h_{f^*(L)} = h_{\phi'} \circ f + O(1) = h_L \circ f + O(1).$$

Thus, we obtain property 2.

Step 3. Next, we consider the case where L is a line bundle in general, and we will define h_L. We note that there are globally generated line bundles L_1, L_2 such that $L = L_1 \otimes L_2^{-1}$. Indeed, if we take any ample line bundle A over X, then the definition of ampleness [11, p. 153] tells us that $L \otimes A^n$ is globally generated for any sufficiently large n. Thus, for a sufficiently large n, if we set $L_1 = L \otimes A^n$, $L_2 = A^n$, then L_1, L_2 are both globally generated, and $L = L_1 \otimes L_2^{-1}$.

Noting that h_{L_1} and h_{L_2} have been defined in Step 1, we define h_L by $h_L = h_{L_1} - h_{L_2}$. We need to check that this definition of h_L is well-defined, i.e., h_L is independent of the choices of L_1 and L_2. So, let M_1 and M_2 be globally generated line bundles such that $L = M_1 \otimes M_2^{-1}$. Our goal is to show that $h_{L_1} - h_{L_2} = h_{M_1} - h_{M_2}$ modulo $B(X)$. But this follows from $L_1 \otimes M_2 = M_1 \otimes L_2$ and the equality $h_{L_1} + h_{M_2} = h_{M_1} + h_{L_2} + O(1)$ in Step 2. Thus, $h_L = h_{L_1} - h_{L_2}$ is well-defined modulo $B(X)$.

Step 4. In Step 3, we have defined h_L for any line bundle. Let us verify the properties 1, 2, and 3.

To verify property 1, we take line bundles A, A', B, B' that are globally generated such that $L_1 = A \otimes A'^{-1}$ and $L_2 = B \otimes B'^{-1}$. Then $A \otimes B$ and $A' \otimes B'$ are globally generated, and $L_1 \otimes L_2 = (A \otimes B) \otimes (A' \otimes B')^{-1}$. Modulo $B(X)$, we have

$$
\begin{aligned}
h_{L_1 \otimes L_2} = h_{(A \otimes B) \otimes (A' \otimes B')^{-1}} &= h_{A \otimes B} - h_{A' \otimes B'} \\
&= (h_A + h_B) - (h_{A'} + h_{B'}) = (h_A - h_{A'}) + (h_B - h_{B'}) \\
&= h_{L_1} + h_{L_2}.
\end{aligned}
$$

To verify property 2, we take line bundles C, C' that are globally generated such that $L = C \otimes C'^{-1}$. Then, modulo $B(X)$, we have

$$
\begin{aligned}
h_{f^*(L)} = h_{f^*(C) \otimes f^*(C')^{-1}} &= h_{f^*(C)} - h_{f^*(C')} \\
&= h_C \circ f - h_{C'} \circ f = h_L \circ f.
\end{aligned}
$$

We have already shown property 3 in Step 1.

Step 5. Finally, the uniqueness is obvious from the construction in Steps 1 and 3. Indeed, Suppose that L is globally generated, and let $\phi_{|L|}$ be a morphism associated to $|L|$. Then, by property 3, h_L must be equal to $h_{\phi_{|L|}}$ modulo $B(X)$.

Suppose that L is a line bundle in general. We take line bundles L_1, L_2 that are globally generated such that $L = L_1 \otimes L_2^{-1}$, Then by 1, h_L must be equal to $h_{L_1} - h_{L_2}$ modulo $B(X)$. $\square$

Proposition 3.10 (positivity of heights)   *Let X be a projective variety over $\overline{\mathbb{Q}}$, and let L be a line bundle over X. Let B be the Zariski closed subset of X defined by the ideal sheaf*

$$
\text{Image}(H^0(X, L) \otimes L^{-1} \to \mathcal{O}_X).
$$

Then there is a constant C such that, for any $x \in (X \setminus B)(\overline{\mathbb{Q}})$, $h_L(x) \geq C$.

Proof   Let s be a nonzero element of $H^0(X, L)$. We first claim the following:

Claim 1   There is a constant C' such that, for any $x \in X(\overline{\mathbb{Q}})$ with $s(x) \neq 0$, we have $h_L(x) \geq C'$.

Proof   First, we take line bundles L_1, L_2 that are globally generated such that $L = L_1 \otimes L_2^{-1}$. Let $\{s_1, \ldots, s_n\}$ be a basis of $H^0(X, L_2)$. Then $\{s s_i\}$ are linearly independent elements of $H^0(X, L_1)$. We take $t_{n+1}, \ldots, t_N \in H^0(X, L_1)$ such that $\{ss_1, \ldots, ss_n, t_{n+1}, \ldots, t_N\}$ gives a basis for $H^0(X, L_1)$. Then, if $x \in X(\overline{\mathbb{Q}})$ is any point with $s(x) \neq 0$, then, modulo $B(X)$, we have

$$
\begin{aligned}
h_L(x) &= h_{L_1}(x) - h_{L_2}(x) \\
&= h(s(x)s_1(x) : \ldots : s(x)s_n(x) : t_{n+1}(x) : \ldots : t_N(x)) \\
&\qquad\qquad\qquad\qquad\qquad\qquad\qquad - h(s_1(x) : \ldots : s_n(x)) \\
&\geq h(s(x)s_1(x) : \ldots : s(x)s_n(x)) - h(s_1(x) : \ldots : s_n(x)) \\
&= h(s_1(x) : \ldots : s_n(x)) - h(s_1(x) : \ldots : s_n(x)) = 0.
\end{aligned}
$$

Thus, we obtain the claim. $\qquad\qquad\qquad\qquad\qquad\qquad\qquad\qquad\qquad\qquad\square$

We resume the proof of Proposition 3.10. Let $\{s_1, \ldots, s_n\}$ be a basis of $H^0(X, L)$. By the above claim, there is a constant C_i such that, if $x \in X(\overline{\mathbb{Q}})$ satisfies $s_i(x) \neq 0$, then $h_L(x) \geq C_i$. We set

$$
C = \min\{C_1, \ldots, C_n\}.
$$

Since $B = \{x \in X \mid s_1(x) = \cdots = s_n(x) = 0\}$, we have $h_L(x) \geq C$ for any $x \in (X \setminus B)(\overline{\mathbb{Q}})$. $\qquad\qquad\qquad\qquad\qquad\qquad\qquad\qquad\qquad\qquad\square$

3.5 Northcott's Finiteness Theorem

In this section, we prove Northcott's finiteness theorem. This theorem gives an important criterion on finiteness of rational points.

We begin by showing some lemmas. Recall from Section 3.3 that $h^+(x) = h(1 : x)$ for $x \in \overline{\mathbb{Q}}$.

Lemma 3.11 *Let $x \in \overline{\mathbb{Q}}$. If $y \in \overline{\mathbb{Q}}$ is conjugate to x over $\mathbb{Q}$, then $h^+(y) = h^+(x)$.*

Proof We take a number field K with $x \in K$. By Proposition 3.5, $h^+(x)$ in independent of the choice of K, so we may assume that $K/\mathbb{Q}$ is a Galois extension. Then there exists $\tau \in \operatorname{Gal}(K/\mathbb{Q})$ such that $y = \tau(x)$. For a prime $p \in \mathbb{Z}$, let $\{P_1, \ldots, P_r\}$ be the prime ideals of O_K over p. Then τ gives a permutation of $\{P_1, \ldots, P_r\}$ (see, for example, [23, Chapter I, Proposition (9.1)]). Likewise, τ gives a permutation of $K(\mathbb{C})$. Then the definition of h^+ gives $h^+(\tau(x)) = h^+(x)$. $\qquad\qquad\qquad\qquad\square$

Lemma 3.12 *Let $x_1, \ldots, x_n \in \overline{\mathbb{Q}}$. Then we have the following:*

1. $h^+\left(\prod_{i=1}^n x_i\right) \leq \sum_{i=1}^n h^+(x_i).$

2. $h^+ \left(\sum_{i=1}^n x_i \right) \le h(1 : x_1 : \cdots : x_n) + \log(n)$.

3. $h(1 : x_1 : \cdots : x_n) \le \sum_{i=1}^n h^+(x_i)$.

Proof We take a number field K with $x_1, \ldots, x_n \in K$. For any $v \in M_K$, we have

$$\max \left\{ 1, \left| \prod_{i=1}^n x_i \right|_v \right\} \le \prod_{i=1}^n \max \{ 1, |x_i|_v \}.$$

Taking the logarithm and taking the sum over $v \in M_K$, we obtain the inequality 1.

The inequality 2 follows from

$$\max \left\{ 1, \left| \sum_{i=1}^n x_i \right|_v \right\} \le \begin{cases} \max \{ 1, |x_1|_v, \ldots, |x_n|_v \} & (\text{if } v \in M_K^{\text{fin}}), \\ n \max \{ 1, |x_1|_v, \ldots, |x_n|_v \} & (\text{if } v \in M_K^\infty). \end{cases}$$

To show 3, we note that $\max\{1, |x_1|_v, \cdots, |x_n|_v\} \le \prod_{i=1}^n \max\{1, |x_i|_v\}$. Then taking the logarithm and taking the sum over $v \in M_K$, we obtain the inequality 3. $\qquad \square$

Next, we prove the following lemma that treats the one-dimensional case of Northcott's finiteness theorem:

Lemma 3.13 *For any integer $d \ge 1$ and real number $M \ge 0$, the set*

$$\left\{ x \in \overline{\mathbb{Q}} \mid h^+(x) \le M, \quad [\mathbb{Q}(x) : \mathbb{Q}] \le d \right\}$$

is finite.

Proof First we suppose that $d = 1$. This means that $x \in \mathbb{Q}$. We write $x = p/q$ with $p, q \in \mathbb{Z}$, $\mathrm{GCD}(p, q) = 1$. Then $h^+(x) = h(p, q) = \log \max\{|p|, |q|\}$. Thus, we have $|p| \le e^M$ and $|q| \le e^M$, and we obtain the assertion.

Next, we consider the general case. For $x \in \overline{\mathbb{Q}}$, let $P(T)$ be the minimal polynomial of x over $\mathbb{Q}$. We set $n = \deg(P(T))$. We denote the roots of $P(T)$ by $x_1, \ldots, x_n$ with $x_1 = x$. We set

$$P(T) = (T - x_1) \cdots (T - x_n) = T^n + \sum_{i=1}^n (-1)^i s_i T^{n-i},$$

where $s_i = \sum_{1 \le j_1 < \cdots < j_i \le n} x_{j_1} \cdots x_{j_i}$ is the i-th symmetric polynomial of $x_1, \ldots, x_n$.

It follows from Lemma 3.11 and Lemma 3.12 that

$$
\begin{aligned}
h^+(s_i) &\leq \sum_{1 \leq j_1 < \cdots < j_i \leq n} h^+\left(x_{j_1} \cdots x_{j_i}\right) + \log \binom{n}{i} \\
&\leq \sum_{1 \leq j_1 < \cdots < j_i \leq n} \left(h^+(x_{j_1}) + \cdots + h^+(x_{j_i})\right) + \log \binom{n}{i} \\
&\leq i \binom{n}{i} h^+(x) + \log \binom{n}{i} \leq n2^n h^+(x) + n \log(2).
\end{aligned}
$$

Suppose that $x \in \overline{\mathbb{Q}}$ satisfies $h^+(x) \leq M$ and $[\mathbb{Q}(x) : \mathbb{Q}] \leq d$. Then $n \leq d$, and we obtain $h^+(s_i) \leq d \log(2) + d2^d M$. Each s_i is a rational number, so the case of $d = 1$ implies that there are only finitely many such s_i. Further, the degree of P is at most d, so there are only finitely many such P. Since x is a root of P, we conclude that there are only finitely many $x \in \overline{\mathbb{Q}}$ with $h^+(x) \leq M$ and $[\mathbb{Q}(x) : \mathbb{Q}] \leq d$. $\qquad\square$

We deduce Northcott's finiteness theorem from Lemma 3.13. Before we state the theorem, we recall the field of definition of a point $x \in \mathbb{P}^n(\overline{\mathbb{Q}})$. We write $x = (x_0 : \cdots : x_n)$. Take i with $x_i \neq 0$, and we set

$$
\mathbb{Q}(x) = \mathbb{Q}(x_0/x_i, \ldots, x_n/x_i).
$$

If j satisfies $x_j \neq 0$, then $x_k/x_j = x_k/x_i(x_j/x_i)^{-1}$, and we have

$$
\mathbb{Q}(x_0/x_i, \ldots, x_n/x_i) = \mathbb{Q}(x_0/x_j, \ldots, x_n/x_j).
$$

We see that $\mathbb{Q}(x)$ depends only on x. We call $\mathbb{Q}(x)$ the field of definition of x.

Theorem 3.14 (Northcott's finiteness theorem) *Let X be a projective variety over $\overline{\mathbb{Q}}$, and let L be an ample line bundle over X. Then for any $d \geq 1$ and $M \geq 0$, the set*

$$
\left\{ x \in X(\overline{\mathbb{Q}}) \mid h_L(x) \leq M, \quad [\mathbb{Q}(x) : \mathbb{Q}] \leq d \right\}
$$

is finite.

Proof By taking a sufficiently large m, and replacing L by $L^{\otimes m}$, we may assume that L is very ample (see Theorem 3.9, 1). Then X is embedded into projective space by the linear system $|L|$, so we may assume that X is projective space $\mathbb{P}^n$. In other words, it suffices to show that

$$\left\{ x = (x_0, \dots, x_n) \in \mathbb{P}^n(\overline{\mathbb{Q}}) \mid h(x) \leq M, \quad [\mathbb{Q}(x) : \mathbb{Q}] \leq d \right\}$$

is a finite set. We set

$$U_i = \left\{ x = (x_0, \dots, x_n) \in \mathbb{P}^n(\overline{\mathbb{Q}}) \mid h(x) \leq M, \ [\mathbb{Q}(x) : \mathbb{Q}] \leq d, \ x_i \neq 0 \right\}.$$

Since $\bigcup_{i=0}^n U_i$ is equal to the previous set, if suffices to show that each U_i is a finite set. We are going to show that U_0 is a finite set. The other cases are similar. We put $y_i = x_i/x_0$. Then we have

$$h(x) = h(1 : y_1 : \dots : y_n) \geq h^+(y_i), \qquad \mathbb{Q}(y_i) \subseteq \mathbb{Q}(x)$$

for all i. By Lemma 3.13, there are only finitely many such y_i, and thus, there are only finitely many such x. $\qquad\qquad\square$

As an application of Northcott's finiteness theorem, we show the following lemma. Assertion 1 is called Kronecker's theorem.

Lemma 3.15 *Let K be a number field and let $h \colon \mathbb{P}^n(K) \to \mathbb{R}$ be the absolute (logarithmic) Weil height function. Then we have the following assertions:*

1. For $x = (x_0 : \dots : x_n) \in \mathbb{P}^n(K)$, we have $h(x) \geq 0$. Further $h(x) = 0$ if and only if

$$(x_0, \dots, x_n) = \lambda(y_0, \dots, y_n),$$

where $\lambda \in K^\times$ and y_i is 0 or a root of unity.
2. The subset $h(\mathbb{P}^n(K))$ is discrete in $\mathbb{R}$.

Proof 1. The first assertion was shown in Proposition 3.5,3. To see the second assertion, suppose that every y_i is 0 or a root of unity. Then by the definition of $h(x)$, we have $h(x) = 0$. Now we assume that $h(x) = 0$. Without loss of generality, we may assume that $x_0 \neq 0$. We put $y_i = x_i/x_0$ for $i = 1, \dots, n$. Then, for each i, we have

$$0 \leq h^+(y_i) \leq h^+(y_1, \dots, y_n) = h(x) = 0,$$

and we obtain $h^+(y_i) = 0$. Suppose that $y_i \neq 0$, and let G be the subgroup of $K^\times$ generated by y_i. By Proposition 3.5, 4, we have $h^+(y_i^n) = |n|h^+(y_i) = 0$. It follows from Northcott's finiteness theorem that G is a finite group. Thus, y_i is a root of unity 1. We obtain Assertion 1.

2. This is obvious from Northcott's finiteness theorem. $\qquad\qquad\square$

3.6 Introduction to Abelian Varieties

In this section, we explain some of the basics of abelian varieties. We refer the reader to Mumford [22] for details. Let F be a field of characteristic zero and let $\overline{F}$ denote an algebraic closure of F.

Let X be an algebraic variety over F, i.e., a reduced and irreducible scheme of finite type over $\mathrm{Spec}(F)$. We say that X is *geometrically irreducible* if the base extension of X to $\overline{F}$ is still irreducible. In this case, $X_E :=$ $X \times_{\mathrm{Spec}(F)} \mathrm{Spec}(E)$ is irreducible for any field extension (not necessarily algebraic) E of F.

In the following, the fiber products of schemes are taken over the base field F unless otherwise specified.

An algebraic variety G over F is called a *group variety* if there are three morphisms

$$m_G \colon G \times G \to G, \quad i_G \colon G \to G, \quad e_G \colon \mathrm{Spec}(F) \to G$$

such that the following three diagrams commute. (These diagrams respectively correspond to the existence of the identity, the existence of the inverse, and the associative law in the definition of a group.)

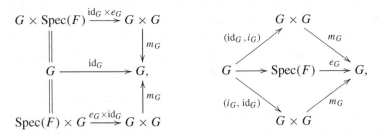

Let G and H be group varieties over F and let $f \colon G \to H$ be a morphism over F. If the diagram

$$
\begin{array}{ccc}
G \times G & \xrightarrow{\;f \times f\;} & H \times H \\
{\scriptstyle m_G}\downarrow & & \downarrow{\scriptstyle m_H} \\
G & \xrightarrow{\quad f \quad} & H
\end{array}
$$

Figure 3.2 Niels Abel.
Source: Archives of the Mathematisches Forschungsinstitut Oberwolfach.

commutes, we say that f is a *homomorphism*. For a homomorphism $f : G \to H$, we denote the fiber product of f and e_H by $\mathrm{Ker}(f) = f^{-1}(e_H)$. Then $\mathrm{Ker}(f)$ becomes a closed subgroup variety of G.

Remark 3.16 In other words, an algebraic variety G is a *group variety* if $G(R)$ (see N7 on p. 164) is a group for any F-algebra R and if, for any homomorphism $R \to R'$ of F-algebras, the induced map $G(R) \to G(R')$ is a group homomorphism. Further, for group varieties G and H, a morphism $f : G \to H$ is called a *homomorphism* if the map $G(R) \to H(R)$ is a group homomorphism for any F-algebra R.

We remark that any group variety is smooth. Indeed, there is a nonempty open subvariety of G that is smooth over F (see [11, Lemma III.10.5]). Then using translations by elements in $G(\overline{F})$, we see that $G_{\overline{F}}$ is smooth at every closed point.

An *abelian variety* (Figure 3.2) over F is a geometrically irreducible, projective group variety A over F. By the above remark, any abelian variety is smooth. Further, anticipating that any abelian variety is commutative as a group (see Theorem 3.18), we use the additive notation $+$, $-$, 0 instead of m_A, i_A, e_A for an abelian variety.

The following lemma is fundamental and very useful:

Lemma 3.17 (Rigidity lemma) *Let X, Y, Z be geometrically irreducible algebraic varieties over F and let $f : X \times Y \to Z$ be a morphism over F. We assume that X is projective over F. If there are points $x_0 \in X(F)$ and $y_0 \in Y(F)$ such that both $\{x_0\} \times Y$ and $X \times \{y_0\}$ are mapped by f to a single point $z_0 \in Z(F)$, then the whole $X \times Y$ is mapped by f to z_0.*

Proof First, let S, T be algebraic varieties over F, and assume that S is projective over F and that T is affine. We claim that, for any morphism $u: S \to T$ over F, the image of u consists of a single closed point. Indeed, u is the composition of the morphism $\mathrm{Spec}(H^0(S, \mathcal{O}_S)) \to T$, which is induced from $u^*: H^0(T, \mathcal{O}_T) \to H^0(S, \mathcal{O}_S)$, and the canonical morphism $S \to \mathrm{Spec}(H^0(S, \mathcal{O}_S))$. Since $H^0(S, \mathcal{O}_S)$ is a finite dimensional integral domain over F (see [11, Theorem II.5.9]), $H^0(S, \mathcal{O}_S)$ is a finite field extension of F. Thus, $\mathrm{Spec}(H^0(S, \mathcal{O}_S))$ is a single point, and we obtain the claim.

Next, let $p: X \times Y \to Y$ denote the second projection, let U be an affine open neighborhood of z_0, and let $V = p(f^{-1}(Z \setminus U))$. Since X is projective, p is a closed morphism and $Y \setminus V$ is an open subset of Y. Further, $y_0 \in (Y \setminus V)(F)$, so $Y \setminus V$ is nonempty. For any closed point y in $Y \setminus V$, $X \times \{y\}$ is projective over F and is mapped by f to the affine variety U. Then it follows from the above claim that f is a constant morphism on $X \times \{y\}$ and $f(X \times \{y\}) = \{f(x_0, y)\} = \{z_0\}$. Since $X \times Y$ is irreducible and the Zariski closed subset $f^{-1}(z_0)$ contains the open subset $X \times (Y \setminus V)$, we have $f^{-1}(z_0) = X \times Y$. $\qquad\square$

Suppose that A and B are abelian varieties over F and that $f: A \to B$ is a morphism over F with $f(0) = 0$. As a first application of the rigidity lemma, we show that f is a homomorphism. Indeed, the morphism $\phi: A \times A \to A$ defined by $\phi(a_1, a_2) = f(a_1 + a_2) - f(a_2) - f(a_1)$ satisfies $\phi(A \times \{0\}) = \phi(\{0\} \times A) = \{0\}$, so, by the rigidity lemma, ϕ is the zero morphism. It follows that $f(a_1 + a_2) = f(a_1) + f(a_2)$ for any $a_1, a_2 \in A(\overline{F})$.

Theorem 3.18 *Any abelian variety A is commutative as a group.*

Proof The inverse morphism $i_A: A \to A$ satisfies $i_A(0) = 0$, so the above argument shows that i_A is a homomorphism. Then, for any $a_1, a_2 \in A(\overline{F})$,

$$-((-a_1) + (-a_2)) = -(-a_1) + (-(-a_2)) = a_1 + a_2.$$

On the other hand, the group laws tell us that $-((-a_1) + (-a_2)) = a_2 + a_1$. Then $a_1 + a_2 = a_2 + a_1$ for any $a_1, a_2 \in A(\overline{F})$. $\qquad\square$

Lemma 3.19 (Seesaw theorem) *Let X and T be geometrically irreducible algebraic varieties over F and let L be a line bundle on $X \times T$. We assume that X is projective over F. Then the subset,*

$$T_1 = \left\{ t \in T \mid L|_{X \times \{t\}} \text{ is trivial} \right\}$$

is Zariski closed in T. Further, if we endow T_1 with the reduced induced scheme structure and denote by $p: X \times T_1 \to T_1$ the second projection, then there is a line bundle M on T_1 such that $L|_{X \times T_1} \cong p^* M$.

Proof First, we note that a line bundle L on a projective variety X is trivial if and only if $\dim_F(H^0(X, L)) \geq 1$ and $\dim_F(H^0(X, L^{-1})) \geq 1$. Indeed, a nonzero section in $H^0(X, L)$ (resp. $H^0(X, L^{-1})$) defines a homomorphism $u \colon \mathcal{O}_X \to L$ (resp. $v \colon L \to \mathcal{O}_X$). The composition $v \circ u \colon \mathcal{O}_X \to \mathcal{O}_X$ maps 1 to a nonzero element in $H^0(X, \mathcal{O}_X)$. Since $H^0(X, \mathcal{O}_X)$ is a field, both u and v are isomorphisms.

It follows that

$$T_1 = \left\{ t \in T \; \middle| \; \begin{array}{l} \dim_{\kappa(t)}(H^0(X \times \{t\}, L|_{X \times \{t\}})) \geq 1 \text{ and} \\ \dim_{\kappa(t)}(H^0(X \times \{t\}, L^{-1}|_{X \times \{t\}})) \geq 1 \end{array} \right\}.$$

The right-hand side is a Zariski closed subset of T by the upper semicontinuity of the dimensions of the cohomology groups (see [11, Theorem III.12.8]), so we obtain the first assertion.

Next, since L is trivial on each fiber $X \times \{t\}$ of $X \times T_1 \to T_1$, we have $\dim_{\kappa(t)} H^0(X \times \{t\}, L|_{X \times \{t\}}) = 1$. It follows from Grauert's theorem [11, Corollary III.12.9] that $M := p_* L$ is a line bundle on T_1. As the natural homomorphism $p^* M \to L$ coincides on each fiber with the natural homomorphism

$$H^0(X \times \{t\}, L|_{X \times \{t\}}) \otimes \mathcal{O}_{X \times \{t\}} \to L|_{X \times \{t\}},$$

and $L|_{X \times \{t\}}$ is trivial for each t, we obtain that $p^* M \cong L$ (see [11, Proposition II.1.1]). □

Suppose that L_1 and L_2 are line bundles on $X \times T$. The seesaw theorem (Lemma 3.19) tells us that $L_1 \otimes L_2^{-1} \in p^*(\mathrm{Pic}(T))$ if and only if

$$L_1|_{X \times \{t\}} \cong L_2|_{X \times \{t\}}$$

for any $t \in T$. Further, if there is an $x \in X(F)$ with $L_1|_{\{x\} \times T} \cong L_2|_{\{x\} \times T}$, then L_1 and L_2 are isomorphic. Indeed, we have $L_1 \otimes L_2^{-1} \cong p^* M$ for some line bundle M on T, and the restriction $p^* M|_{\{x\} \times T}$ is trivial, so M is trivial.

Remark 3.20 We set $X = \mathbb{P}^n$, $Y = \mathbb{P}^m$, and $Z = \mathbb{P}^n \times \mathbb{P}^m$. We denote by $p_1 \colon Z \to X$ and $p_2 \colon Z \to Y$ the natural projections. Using the seesaw theorem (Lemma 3.19), let us show that the homomorphism $\iota \colon \mathbb{Z} \times \mathbb{Z} \to \mathrm{Pic}(Z)$ defined by $(a, b) \mapsto \mathcal{O}_Z(a, b) := p_1^* \mathcal{O}_X(a) \otimes p_2^* \mathcal{O}_Y(b)$ is an isomorphism. Indeed, the injectivity of ι follows from $\deg(\mathcal{O}_Z(a, b)|_{X \times \{y_0\}}) = a$, $\deg(\mathcal{O}_Z(a, b)|_{\{x_0\} \times Y}) = b$ for any $x_0 \in X(F)$ and $y_0 \in Y(F)$. To show the surjectivity, we take any line bundle L on Z. Noting that $\deg(L|_{X \times \{y\}})$ is constant for each y in Y (see [11, Theorem III.9.9]), we denote this value by a. Since $\deg(L|_{X \times \{y\}}) = \deg(p_1^* \mathcal{O}_X(a)|_{X \times \{y\}})$, i.e., $L|_{X \times \{y\}} \cong p_1^* \mathcal{O}_X(a)|_{X \times \{y\}}$, the seesaw theorem (Lemma 3.19) implies that there is an integer b such that $L \cong p_1^* \mathcal{O}(a) \otimes p_2^* \mathcal{O}(b)$ (see also [11, Corollary II.6.17]). Thus, ι is surjective.

To show further properties of abelian varieties, we first note the following:

Lemma 3.21 *Let X be a projective variety, and let x_0 and x_1 be closed points of X. Then there is a projective curve C on X passing through x_0 and x_1.*

Proof If $x_0 = x_1$, then the result is obvious, so we assume that $x_0 \neq x_1$. We show the lemma by induction on the dimension of X. The case of $\dim X = 1$ is obvious. Suppose $\dim X \geq 2$ and let $\pi : \widetilde{X} \to X$ be the blowup of X along the two points x_0, x_1. We embed $\widetilde{X}$ into projective space $\mathbb{P}^N$. We take N to be the minimum among such embeddings. For a general hyperplane $H \subseteq \mathbb{P}^N$, $\widetilde{Y} = \widetilde{X} \cap H$ is irreducible and is not contained in $\pi^{-1}(x_i)$ (see [11, Theorem II.8.18, Remark III.7.9.1]). Further, since $\dim H + \dim \pi^{-1}(x_i) \geq N$, $H \cap \pi^{-1}(x_i) \neq \emptyset$ (see [11, Theorem I.7.2]). If we set $Y = \pi(\widetilde{Y}) \subseteq X$, then Y is a $(\dim X - 1)$-dimensional subvariety of X that contains x_0 and x_1. By the induction hypothesis, there is a projective curve C on Y that passes through x_0 and x_1. This C is a desired curve. $\qquad\square$

A quadratic function $f(x) = ax^2 + bx$ with no constant term satisfies the equations of the following forms:

$$f(x + y + z) = f(x + y) + f(y + z) + f(z + x) - f(x) - f(y) - f(z),$$
$$f(nx) = \frac{n(n + 1)}{2} f(x) + \frac{n(n - 1)}{2} f(-x).$$

The next theorem, which is the main result of this section, shows that line bundles on abelian varieties satisfy such "quadratic" relations.

Theorem 3.22 (Theorem of the cube I) *Let X be a projective variety over F, let A be an abelian variety over F, and let $f, g, h \colon X \to A$ be morphisms over F. Then, for any line bundle L on A,*

$$(f + g + h)^*(L) \otimes (f + g)^*(L^{-1}) \otimes (g + h)^*(L^{-1}) \otimes (h + f)^*(L^{-1})$$
$$\otimes f^*(L) \otimes g^*(L) \otimes h^*(L) \cong \mathcal{O}_X.$$

Theorem 3.22 is a consequence of the following more general theorem:

Theorem 3.23 (Theorem of the cube II) *Let X, Y, and Z be geometrically irreducible algebraic varieties over F, let $x_0 \in X(F)$, $y_0 \in Y(F)$, and $z_0 \in Z(F)$ be closed points, and let L be a line bundle on $X \times Y \times Z$. Assume that X and Y are projective over F. Then L is trivial if and only if*

$$L|_{\{x_0\} \times Y \times Z}, \quad L|_{X \times \{y_0\} \times Z}, \quad and \quad L|_{X \times Y \times \{z_0\}}$$

are all trivial.

Proof of Theorem 3.23 ⇒ Theorem 3.22   Let $m_{123}: A \times A \times A \to A$ be the morphism defined by $(a_1, a_2, a_3) \mapsto a_1 + a_2 + a_3$, let $m_{ij}: A \times A \times A \to A$ be the morphism defined by $(a_1, a_2, a_3) \mapsto a_i + a_j$, and let $p_i: A \times A \times A \to A$ be the i-th projection. We define the line bundle M on $A \times A \times A$ to be

$$m_{123}^*(L) \otimes m_{12}^*(L^{-1}) \otimes m_{23}^*(L^{-1}) \otimes m_{31}^*(L^{-1}) \otimes p_1^*(L) \otimes p_2^*(L) \otimes p_3^*(L).$$

Then M is trivial on $\{0\} \times A \times A$, $A \times \{0\} \times A$, and $A \times A \times \{0\}$, respectively, so M itself is trivial by Theorem 3.23. Pulling back M by the morphism $(f, g, h): X \to A \times A \times A$, we obtain Theorem 3.22.   □

Proof of Theorem 3.23   The "only if" part is obvious. We are going to show the "if" part. We divide the proof into three steps.

Step 1. Let p_{13} denote the projection $X \times Y \times Z \to X \times Z$ to the first and the last factor. If we show that $L|_{\{x\} \times Y \times \{z\}}$ is trivial for every closed points $x \in X$ and $z \in Z$, then, by the seesaw theorem (Lemma 3.19), we can find a line bundle M on $X \times Z$ such that $p_{13}^* M \cong L$. Further, $L|_{X \times \{y_0\} \times Z}$ is trivial by the hypothesis, so M is trivial. Thus, L is trivial and we will obtain the theorem.

Step 2. For any closed point $x \in X$, Lemma 3.21 tells us that there is a projective curve C on X passing x_0 and x. Let $\widetilde{C} \to C$ be the normalization of C. Suppose that we show Theorem 3.23 for $X = \widetilde{C}$. Then $L|_{\widetilde{C} \times Y \times Z}$ will be trivial, so $L|_{\{x\} \times Y \times \{z\}}$ will be trivial for any closed point $z \in Z$. Since x is an arbitrary closed point of X, by Step 1, we will obtain Theorem 3.23 for general X.

Step 3. By Step 2, we may assume that X is a smooth projective curve. To conclude the proof, we here use some properties of the Jacobian variety J of X, which we explain in Section 3.8 below. By [11, Theorem III.9.9], L has degree zero on each fiber of $X \times (Y \times Z) \to Y \times Z$. Thus, L defines a morphism

$$f: Y \times Z \to J.$$

Since both $\{y_0\} \times Z$ and $Y \times \{z_0\}$ are mapped to 0 by f, the rigidity lemma (Lemma 3.17) tells us that the whole $Y \times Z$ is also mapped to 0 by f. Thus, L is trivial on $X \times Y \times Z$.   □

For an integer n, we define the morphism $[n]_A: A \to A$ as follows. If $n = 0$, then $[n]_A$ is the composition $A \to \mathrm{Spec}(F) \xrightarrow{e_A} A$. If $n > 0$, then

$$[n]_A: A \xrightarrow{\text{diag.}} \overbrace{A \times \cdots \times A}^{n \text{ times}} \xrightarrow{\text{mult.}} A,$$

where diag. is the diagonal morphism, and mult. is the multiplication morphism. If $n < 0$, then $[n]_A = i_A \circ [|n|]_A$. To put it simply, $[n]_A: A \to A$ is the multiplication-by-n morphism on A defined by $x \mapsto nx$. In the following, we simply denote $[n]_A$ by $[n]$ when no confusion is likely.

Corollary 3.24   *For any integer n and a line bundle L on A,*

$$[n]^*L \cong L^{\frac{n(n+1)}{2}} \otimes [-1]^*L^{\frac{n(n-1)}{2}}.$$

Proof   If $n = 0, \pm 1$, then the assertion is obvious. We are going to show the result by induction on n. If we apply the theorem of the cube (Theorem 3.22) to $f = [n]$, $g = [1]$, and $h = [-1]$, then

$$[n]^*L \otimes [n+1]^*L^{-1} \otimes [n-1]^*L^{-1} \otimes [0]^*L^{-1} \otimes [n]^*L \otimes [1]^*L \otimes [-1]^*L \cong \mathcal{O}_A.$$

If $n \geq 1$, then by induction hypothesis,

$$[n+1]^*L$$

$$\cong [n]^*L^2 \otimes [n-1]^*L^{-1} \otimes L \otimes [-1]^*L$$

$$\cong \left(L^{n(n+1)} \otimes [-1]^*L^{n(n-1)} \right) \otimes \left(L^{-\frac{n(n-1)}{2}} \otimes [-1]^*L^{-\frac{(n-1)(n-2)}{2}} \right)$$

$$\otimes L \otimes [-1]^*L$$

$$\cong L^{\frac{(n+1)(n+2)}{2}} \otimes [-1]^*L^{\frac{n(n+1)}{2}}.$$

We can show the case of $n < -1$ similarly.   □

A line bundle L on an abelian variety A is said to be *even* (resp. *odd*) if $[-1]^*L \cong L$ (resp. $[-1]^*L \cong L^{-1}$).[1] By Corollary 3.24, we have

$$[n]^*L \cong \begin{cases} L^{n^2} & \text{(if L is even),} \\ L^n & \text{(if L is odd).} \end{cases} \tag{3.11}$$

Note that, there always exists an even and ample line bundle on A. Indeed, we take any ample line bundle L on A. Then $[-1]^*L$ is ample, so $L \otimes [-1]^*L$ is even and ample.

Corollary 3.25   *Let A be an abelian variety of dimension g over F.*

1. *For any integer n, $[n] \colon A \to A$ is a finite and flat morphism of degree n^{2g}.*
2. *The abelian group $A(\overline{F})$ is divisible, i.e., for any $x \in A(\overline{F})$ and for any positive integer n, there is $y \in A(\overline{F})$ with $[n](y) = x$.*

Proof   1. Let L be an even and ample line bundle on A. Then, by Corollary 3.24, $[n]^*L$ is ample. On the other hand, $[n]^*L$ is trivial on $\mathrm{Ker}([n])$. Since the trivial line bundle on $\mathrm{Ker}([n])$ gives an embedding into a projective space,

---

[1] An even line bundle is also called a *symmetric* line bundle (see, e.g., [22]).

we have $\dim \operatorname{Ker}([n]) = 0$. Further, since $[n]$ is proper, $[n]$ is finite. It follows from

$$(([n]^*L)^{\cdot g}) = ((L^{n^2})^{\cdot g}) = n^{2g}(L^{\cdot g})$$

that $[n]$ has degree n^{2g}. Since $[n]$ has equidimensional fibers, $[n]$ is flat (see, e.g., [18, p. 179, Corollary]).

2. The image $[n](A)$ of $[n]$ is a closed subvariety of A and $\dim A = \dim[n](A)$ by 1. Thus, $[n]$ is surjective, and we obtain the assertion. $\square$

For $a \in A(\overline{F})$, we define the morphism $T_a : A_{\overline{F}} \to A_{\overline{F}}$ by

$$T_a : x \mapsto x + a.$$

Let $\operatorname{Pic}(A_{\overline{F}})$ be the Picard group of $A_{\overline{F}}$ (see N10 on p. 164). For any given line bundle L on $A_{\overline{F}}$, we define the map $\lambda_L : A(\overline{F}) \to \operatorname{Pic}(A_{\overline{F}})$ by

$$\lambda_L : x \mapsto T_x^*L \otimes L^{-1}.$$

Corollary 3.26 (Theorem of the square) *As groups, the map λ_L is a homomorphism from $A(\overline{F})$ to $\operatorname{Pic}(A_{\overline{F}})$.*

Proof Obviously, we may assume $F = \overline{F}$. Let $a, b \in A(F)$. We set $f : A \to \operatorname{Spec}(F) \overset{a}{\to} A$, $g : A \to \operatorname{Spec}(F) \overset{b}{\to} A$, and $h = \operatorname{id}_A : A \to A$. Then $f + g + h = T_{a+b}$. Applying the theorem of the cube (Theorem 3.22) to f, g, h, we obtain

$$T_{a+b}^*L \otimes T_a^*L^{-1} \otimes T_b^*L^{-1} \otimes L \cong \mathcal{O}_A.$$

Thus, $\lambda_L(a + b) = \lambda_L(a) + \lambda_L(b)$. $\square$

Remark 3.27 If D is an effective Cartier divisor on an abelian variety A, then $|2D|$ is base point free. In particular, D is nef. Indeed, identifying D with the corresponding Weil divisor, we write $D + a$ and $D - a$ for $T_a(D)$ and $T_{-a}(D)$, respectively. For any $x \in A(\overline{F})$, we choose a point $a \notin \operatorname{Supp}(D - x) \cup \operatorname{Supp}(D + x)$. Then $x \notin \operatorname{Supp}(D - a) \cup \operatorname{Supp}(D + a)$. Further, the theorem of the square (Corollary 3.26) implies that $(D - a) + (D + a) \sim 2D$. Thus, $|2D|$ is base point free.

Corollary 3.28 *Let A be an abelian variety, and let D be an effective Cartier divisor on A. We set $L = \mathcal{O}_A(D)$. (In particular, $|2D|$ is base point free by Remark 3.27.) Then the following are equivalent:*

1. $\operatorname{Ker}(\lambda_L)$ is a finite subgroup of $A(\overline{F})$.

2. Let $\Phi: A \to \mathbb{P}(H^0(A, L^2))$ be a morphism associated to the complete linear system $|2D|$. Then Φ is a finite morphism onto its image.
3. L is ample.

Proof Properties 2 and 3 hold over F if and only if they do over $\overline{F}$. We may assume that $F = \overline{F}$.

$1 \Rightarrow 2$: Suppose that Φ maps a projective curve C on A to a single point. We are going to deduce a contradiction by showing that $\mathrm{Ker}(\lambda_L)$ contains $C - C = \{x_2 - x_1 \mid x_1, x_2 \in C(F)\}$. We write $D = \sum_{i=1}^r a_i D_i$ with $a_i > 0$ and prime divisors D_i on A.

We claim that either $(C + x) \cap D_i = \emptyset$ or $C + x \subseteq D_i$ for any i and for any $x \in A(F)$. Since D_i is an effective Cartier divisor on an abelian variety, it is nef by Remark 3.27. In particular, $(D_i \cdot C) \geq 0$. It follows from

$$0 = (D \cdot C) = a_1 (D_1 \cdot C) + \cdots + a_r (D_r \cdot C)$$

that $(D_i \cdot C) = 0$. Further, since $C + x$ is algebraically equivalent to C, we have $(D_i \cdot C + x) = (D_i \cdot C) = 0$. Thus, we obtain the claim.

Next, we claim that $D_i = D_i + x_1 - x_2$ for any $x_1, x_2 \in C(F)$. Indeed, if $y \in D_i$ is a closed point, then both $C - x_1 + y$ and D_i contain y. Thus, $C - x_1 + y \subseteq D_i$ by the above argument. It follows that $x_2 - x_1 + y \in D_i$, so $y \in D_i + x_1 - x_2$. Thus, $D_i \subseteq D_i + x_1 - x_2$. By switching x_1 and x_2 in the above argument, we have $D_i \supseteq D_i + x_1 - x_2$. Hence, we obtain the claim.

We set $L_i = \mathcal{O}_A(D_i)$. Then $L = L_1^{\otimes a_1} \otimes \cdots \otimes L_r^{\otimes a_r}$. By the above claim, we have $T_{x_2 - x_1}^*(L_i) = L_i$ for each i. It follows that

$$T_{x_2-x_1}^*(L) = T_{x_2-x_1}^*(L_1)^{\otimes a_1} \otimes \cdots \otimes T_{x_2-x_1}^*(L_r)^{\otimes a_r}$$
$$= L_1^{\otimes a_1} \otimes \cdots \otimes L_r^{\otimes a_r} = L.$$

Thus, $x_2 - x_1 \in \mathrm{Ker}(\lambda_L)$, so $C - C \subseteq \mathrm{Ker}(\lambda_L)$. This is a contradiction.

$2 \Rightarrow 3$: Let E be any coherent $\mathcal{O}_A$-module on A. We are going to show that $E \otimes L^{2n}$ is globally generated for any sufficiently large $2n$. We write X for the image $\Phi(A)$ of Φ. Since $\Phi_*(E)$ is coherent on X and $\mathcal{O}_X(1)$ is ample, there is $n_0 > 0$ such that

$$H^0(X, \Phi_*(E) \otimes \mathcal{O}_X(n)) \otimes \mathcal{O}_X \to \Phi_*(E) \otimes \mathcal{O}_X(n)$$

is surjective for any $n \geq n_0$ (see [11, p. 153]). Pulling back by Φ, we find that

$$H^0(X, \Phi_*(E) \otimes \mathcal{O}_X(n)) \otimes \mathcal{O}_A \to \Phi^* \Phi_*(E) \otimes \Phi^* \mathcal{O}_X(n) \qquad (3.12)$$

is also surjective.

On the other hand, it follows from $\Phi^* \mathcal{O}_X(n) \cong L^{2n}$ and the projection formula that $\Phi_*(E \otimes L^{2n}) \cong \Phi_*(E) \otimes \mathcal{O}_X(n)$. Further, since Φ is a finite morphism, the canonical morphism $\Phi^*\Phi_*(E) \to E$ is surjective. Thus, the surjectivity of (3.12) implies the surjectivity of

$$H^0(A, E \otimes L^{2n}) \otimes \mathcal{O}_A \to E \otimes L^{2n}$$

for every $n \geq n_0$. Thus, L^2 is ample, so L is ample.

$3 \Rightarrow 1$: Let $p_i \colon A \times A \to A$ denote the i-th projection. First, we note the equality

$$\mathrm{Ker}(\lambda_L) = \left\{ x \in A(F) \ \middle| \ (m_A^* L^{-1} \otimes p_1^* L)|_{A \times \{x\}} \text{ is trivial} \right\}.$$

In particular, the seesaw theorem (Lemma 3.19) tells us that $\mathrm{Ker}(\lambda_L)$ endowed with the reduced induced scheme structure is regarded as a closed subgroup scheme of A. We denote by B the connected component of $\mathrm{Ker}(\lambda_L)$ containing the identity. Then B is an abelian subvariety of A. We are going to show that $\dim B = 0$.

We set $L' := (m_A^* L^{-1} \otimes p_1^* L \otimes p_2^* L)|_{B \times B}$ on $B \times B$. Since $L'|_{B \times \{x\}} = (m_A^* L^{-1} \otimes p_1^* L)|_{B \times \{x\}} \otimes p_2^* L|_{B \times \{x\}}$ is trivial for any $x \in B$ and $L'|_{\{0\} \times B}$ is trivial, the seesaw theorem (Lemma 3.19) implies that L' is trivial. Pulling back L' by $([1]_B, [-1]_B) \colon B \to B \times B$, we obtain that $L \otimes [-1]_A^* L$ is trivial on B. On the other hand, since L is ample on A, $L \otimes [-1]_A^* L$ is ample on A. Thus, $\dim B = 0$, and we conclude that $\mathrm{Ker}(\lambda_L)$ is a finite set. □

3.7 Height Functions on Abelian Varieties

As we explained in Section 3.4, height functions associated to line bundles are in general defined only up to modulo bounded functions. In this section, we will show that, on abelian varieties, we can define height functions associated to line bundles without ambiguity of bounded functions.

To begin with, we observe the following lemma. Recall from Section 3.4 that, for any real-valued functions h_1, h_2 defined on a set S, we write

$$h_1 = h_2 + O(1),$$

if $h_1 - h_2$ is a bounded function on S.

Lemma 3.29 *Let A be an abelian group and let $h \colon A \to \mathbb{R}$ be a function. If*

$$h(x+y+z) - h(x+y) - h(y+z) - h(z+x) + h(x) + h(y) + h(z) = O(1)$$

as functions on $A \times A \times A$, then there are a symmetric bilinear function
$b \colon A \times A \to \mathbb{R}$ *and a linear function* $l \colon A \to \mathbb{R}$ *such that*

$$h(x) = \frac{1}{2} b(x, x) + l(x) + O(1)$$

for any $x \in A$. *Further* b *and* l *are uniquely determined by* h.

Proof We set $\beta(x, y) := h(x + y) - h(x) - h(y)$. Then β is a symmetric function on $A \times A$. Further, it follows from the assumption,

$$\beta(x + y, z) = h(x + y + z) - h(x + y) - h(z), \text{ and}$$
$$\beta(x, z) + \beta(y, z) = h(x + z) + h(y + z) - h(x) - h(y) - 2h(z)$$

that there is a constant C such that

$$|\beta(x + y, z) - \beta(x, z) - \beta(y, z)| \leq C$$

for all $x, y, z \in A$.

Now we use the so-called telescoping argument. For any positive integer n, we have

$$\begin{cases} |\beta(2^{n+1}x, 2^{n+1}y) - 2\beta(2^n x, 2^{n+1}y)| \leq C & \text{and} \\ |\beta(2^n x, 2^{n+1}y) - 2\beta(2^n x, 2^n y)| \leq C, \end{cases}$$

so

$$|\beta(2^{n+1}x, 2^{n+1}y) - 4\beta(2^n x, 2^n y)| \leq 3C.$$

Thus,

$$|4^{-(n+1)}\beta(2^{n+1}x, 2^{n+1}y) - 4^{-n}\beta(2^n x, 2^n y)| \leq 4^{-(n+1)} \cdot 3C,$$

so the sequence $\{4^{-n}\beta(2^n x, 2^n y)\}$ is Cauchy. We denote the limit by $b(x, y)$

$$b(x, y) = \lim_{n \to \infty} 4^{-n}\beta(2^n x, 2^n y).$$

Note that b is symmetric, because β is symmetric.

It follows from

$$|4^{-N}\beta(2^N x, 2^N y) - \beta(x, y)| \leq 3C \sum_{n=1}^{N} 4^{-n} \leq C$$

that $|b - \beta| \leq C$. Further, we have

$$|4^{-n}\beta(2^n(x + y), 2^n z) - 4^{-n}\beta(2^n x, 2^n z) - 4^{-n}\beta(2^n y, 2^n z)| \leq 4^{-n}C.$$

Letting $n \to \infty$, we see that b is bilinear.

Next, we set $\lambda(x) = h(x) - \frac{1}{2}b(x,x)$. Since

$$\lambda(x+y) - \lambda(x) - \lambda(y) = \beta(x,y) - b(x,y),$$

we have

$$|\lambda(x+y) - \lambda(x) - \lambda(y)| \le C.$$

Thus,

$$|2^{-(n+1)}\lambda(2^{n+1}x) - 2^{-n}\lambda(2^n x)| \le 2^{-(n+1)}C,$$

so the sequence $\{2^{-n}\lambda(2^n x)\}$ is Cauchy. We denote the limit by $l(x)$

$$l(x) = \lim_{n \to \infty} 2^{-n}\lambda(2^n x).$$

We have $|l - \lambda| \le \sum_{n=1}^{\infty} 2^{-n}C = C$. Further, since

$$|2^{-n}\lambda(2^n(x+y)) - 2^{-n}\lambda(2^n x) - 2^{-n}\lambda(2^n y)| \le 2^{-n}C,$$

we see that l is linear by letting $n \to \infty$.

In summary, we have the required decomposition

$$h(x) = \frac{1}{2}b(x,x) + l(x) + O(1).$$

Lastly, we show the uniqueness of b and l. Suppose that there are another symmetric and bilinear function b' and another linear function l' such that $h = \frac{1}{2}b' + l' + O(1)$. It follows from $b(x,x) = h(x) + h(-x) + O(1) = b'(x,x) + O(1)$, $2b(x,y) = b(x+y,x+y) - b(x,x) - b(y,y)$, and $2b'(x,y) = b'(x+y,x+y) - b'(x,x) - b'(y,y)$ that $b = b' + O(1)$. Thus, there is a constant C' such that $|b - b'| \le C'$. Then

$$|b(x,y) - b'(x,y)| = 2^{-n}|b(2^n x,y) - b'(2^n x,y)| \le 2^{-n}C',$$

so we obtain $b = b'$ by letting $n \to \infty$. It follows that $l = l' + O(1)$, i.e., there is a constant C'' such that $|l - l'| \le C''$. Similarly,

$$|l(x) - l'(x)| = 2^{-n}|l(2^n x) - l'(2^n x)| \le 2^{-n}C'',$$

and we have $l = l'$ by letting $n \to \infty$. $\qquad\qquad\qquad\square$

Theorem 3.30 *Let A be an abelian variety over $\overline{\mathbb{Q}}$. For any line bundle L on A, there are a symmetric and bilinear function $b_L : A(\overline{\mathbb{Q}}) \times A(\overline{\mathbb{Q}}) \to \mathbb{R}$ and a linear function $l_L : A(\overline{\mathbb{Q}}) \to \mathbb{R}$ such that*

$$h_L(x) = \frac{1}{2}b_L(x,x) + l_L(x) + O(1)$$

for every $x \in A(\overline{\mathbb{Q}})$. Further, b_L and l_L are uniquely determined by L.

Proof   Let $p_i : A \times A \times A \to A$ denote the i-th projection. Since

$$(p_1 + p_2 + p_3)^*(L) \otimes (p_1 + p_2)^*(L^{-1}) \otimes (p_2 + p_3)^*(L^{-1}) \otimes (p_3 + p_1)^*(L^{-1})$$
$$\otimes \, p_1^*(L) \otimes p_2^*(L) \otimes p_3^*(L) \cong \mathcal{O}_A$$

by Theorem 3.22, we have

$$h_L(x + y + z) - h_L(x + y) - h_L(y + z) - h_L(z + x)$$
$$+ h_L(x) + h_L(y) + h_L(z) = O(1).$$

Thus, the theorem follows from Lemma 3.29.   □

We define $\hat{h}_L : A(\overline{\mathbb{Q}}) \to \mathbb{R}$ by

$$\hat{h}_L(x) = \frac{1}{2} b_L(x, x) + l_L(x),$$

which we call the *canonical height* of L or the *Néron–Tate height* of L.

We are going to see some fundamental properties of the canonical heights. Let $[-1] : A \to A$ be the homomorphism defined by $[-1](x) = -x$. Recall from Section 3.6 that a line bundle L on A is *even* if $[-1]^*(L) \cong L$, and *odd* if $[-1]^*(L) \cong L^{-1}$.

Theorem 3.31   *Let A be an abelian variety over $\overline{\mathbb{Q}}$.*

1. *For any line bundles L_1, L_2 on A, we have $\hat{h}_{L_1 \otimes L_2} = \hat{h}_{L_1} + \hat{h}_{L_2}$.*
2. *If $f : A \to B$ is a homomorphism of abelian varieties over $\overline{\mathbb{Q}}$, then $\hat{h}_{f^*(L)} = \hat{h}_L \circ f$.*
3. *If L is an even line bundle on A, then $l_L = 0$ and, if L is an odd line bundle on A, then $b_L = 0$.*

Proof   1. By Theorem 3.9, we have

$$h_{L_1 \otimes L_2} = h_{L_1} + h_{L_2} + O(1) = \frac{1}{2}\left(b_{L_1} + b_{L_2}\right) + \left(l_{L_1} + l_{L_2}\right) + O(1).$$

Then by the uniqueness of b and l in Theorem 3.30, we obtain $b_{L_1 \otimes L_2} = b_{L_1} + b_{L_2}$, and $l_{L_1 \otimes L_2} = l_{L_1} + l_{L_2}$.

2. By Theorem 3.9, we have

$$h_{f^*(L)} = h_L \circ f + O(1) = \hat{h}_L \circ f + O(1)$$
$$= \frac{1}{2} b_L \circ (f \times f) + l_L \circ f + O(1).$$

Since f is a homomorphism of abelian varieties, the function $b_L \circ (f \times f)$ is symmetric and bilinear and $l_L \circ f$ is linear. Then by the uniqueness of b and l in Theorem 3.30, we obtain $b_L \circ (f \times f) = b_{f^*(L)}$ and $l_L \circ f = l_{f^*(L)}$.

3. If $[-1]^*L = L$, then $\hat{h}_L(-x) = \hat{h}_L(x)$ by Assertion 2. It follows that $l_L(-x) = l_L(x)$, so $l_L(x) = 0$. If $[-1]^*L = L^{-1}$, then $\hat{h}_L(-x) = -\hat{h}_L(x)$ by Assertion 2. It follows that $b_L(-x, -x) = -b_L(x,x)$, so $b_L(x,x) = 0$. Since $2b_L(x,y) = b_L(x+y,x+y) - b_L(x,x) - b_L(y,y)$, we have $b_L = 0$. □

Let A be an abelian variety over $\overline{\mathbb{Q}}$ and let L be an even and ample line bundle on A. From now on, we denote $b_L(x, y)$ by $\langle x, y \rangle_L$ and call it the *Néron–Tate height pairing* of L. By Theorem 3.30, we have $\hat{h}_L(x) = \frac{1}{2}\langle x, x \rangle_L$.

We are going to show some properties of the Néron–Tate height pairings. Let $x \in A(\overline{\mathbb{Q}})$ and n a positive integer. We say that x is an *n-torsion point* if $nx = 0$. We say that x is a torsion point if x is an n-torsion point for some positive integer n. We denote the set of all torsion points of A by $A(\overline{\mathbb{Q}})_{\text{tor}}$.

Theorem 3.32 *Let A be an abelian variety over $\overline{\mathbb{Q}}$, let L be an even and ample line bundle on A, and let $\langle , \rangle_L : A(\overline{\mathbb{Q}}) \times A(\overline{\mathbb{Q}}) \to \mathbb{R}$ be the Néron–Tate height pairing of L.*

1. *If $x \in A(\overline{\mathbb{Q}})_{\text{tor}}$, then $\langle x, y \rangle_L = 0$ for any $y \in A(\overline{\mathbb{Q}})$.*
2. *The symmetric bilinear form on $A(\overline{\mathbb{Q}})/A(\overline{\mathbb{Q}})_{\text{tor}} \times A(\overline{\mathbb{Q}})/A(\overline{\mathbb{Q}})_{\text{tor}}$ induced by $\langle , \rangle_L$ is positive definite.*
3. *Let $p_i : A \times A \to A$ denote the i-th projection. We set*
   $P = (p_1 + p_2)^*(L) \otimes p_1^*(L^{-1}) \otimes p_2^*(L^{-1})$. *Then $\hat{h}_P(x, y) = \langle x, y \rangle_L$ on $A(\overline{\mathbb{Q}}) \times A(\overline{\mathbb{Q}})$.*

Proof 1. We take a positive integer n with $nx = 0$. Then

$$n\langle x, y \rangle_L = \langle nx, y \rangle_L = \langle 0, y \rangle_L = 0.$$

Thus, $\langle x, y \rangle_L = 0$.

2. Since L is ample, L^n is globally generated for a sufficiently large n. Then Proposition 3.10 tells us that there is a constant C such that $\hat{h}_L(x) \geq C$. Since

$$m^2 \hat{h}_L(x) = \hat{h}_L(mx) \geq C$$

for any integer m, we see $\hat{h}_L(x) \geq 0$.

To show the positive definiteness, it suffices to show that $\hat{h}_L(x) = 0$ implies $x \in A(\overline{\mathbb{Q}})_{\text{tor}}$. We take a number field K such that A is defined over K and $x \in A(K)$. We set $G = \{mx \mid m \in \mathbb{Z}\}$. Since $\hat{h}_L(y) = m^2 \hat{h}_L(x) = 0$ for every $y = mx \in G$, Northcott's finiteness theorem (Theorem 3.14) implies that G is a finite group. Thus, $x \in A(K)_{\text{tor}}$.

3. Let $\phi : A \to A \times A$ be the diagonal morphism defined by $\phi(x) = (x, x)$. As before, we let $[n] : A \to A$ be the multiplication-by-n morphism. Then

$$\phi^*(P) \cong [2]^*(L) \otimes L^{-2}.$$

Since $[2]^*(L) \cong L^4$ by Equation (3.11), we have $\phi^*(P) \cong L^2$. Thus,

$$\hat{h}_P(x,x) = 2\hat{h}_L(x) = \langle x, x \rangle_L.$$

It remains to show that $\hat{h}_P(x,y) = \langle x, y \rangle_L$. If we define a morphism $\iota \colon A \times A \to A \times A$ by $\iota(x,y) = (y,x)$, then $\iota^*(P) \cong P$. It follows that

$$\hat{h}_P(x,y) = \hat{h}_P(y,x).$$

We claim that

$$\hat{h}_P(x+y,z) = \hat{h}_P(x,z) + \hat{h}_P(y,z).$$

Indeed, we write $a, b, c \colon A \times A \times A \to A \times A$ for the morphisms defined by $a(x,y,z) = (x+y,z)$, $b(x,y,z) = (x,z)$, $c(x,y,z) = (y,z)$, respectively. Let $q_i \colon A \times A \times A \to A$ denote the i-th projection. Then, it follows from

$$(p_1 + p_2) \circ a = q_1 + q_2 + q_3, \quad (p_1 + p_2) \circ b = q_1 + q_3, \quad (p_1 + p_2) \circ c = q_2 + q_3,$$
$$p_1 \circ a = q_1 + q_2, \qquad\qquad p_1 \circ b = q_1, \qquad\qquad p_1 \circ c = q_2,$$
$$p_2 \circ a = q_3, \qquad\qquad p_2 \circ b = q_3, \qquad\qquad p_2 \circ c = q_3,$$

that

$$a^*(P) \otimes b^*(P^{-1}) \otimes c^*(P^{-1})$$
$$= (q_1+q_2+q_3)^*(L) \otimes (q_1+q_2)^*(L^{-1}) \otimes (q_2+q_3)^*(L^{-1}) \otimes (q_3+q_1)^*(L^{-1})$$
$$\otimes q_1^*(L) \otimes q_2^*(L) \otimes q_3^*(L).$$

Then Theorem 3.22 tells us that

$$a^*(P) \otimes b^*(P^{-1}) \otimes c^*(P^{-1}) \cong \mathcal{O}_{A \times A \times A}.$$

Thus,

$$\hat{h}_P(x+y,z) = \hat{h}_P(x,z) + \hat{h}_P(y,z),$$

and we obtain the claim. Expanding

$$\hat{h}_P(x+y,x+y) = \langle x+y, x+y \rangle_L,$$

we get $\hat{h}_P(x,y) = \langle x, y \rangle_L$. $\qquad\qquad\qquad\qquad\qquad\qquad\qquad\square$

In fact, we can strengthen Theorem 3.32,2 as follows:

Proposition 3.33 *We use the same notation as in Theorem 3.32. Then the Néron–Tate height pairing $\langle\,,\,\rangle_L \colon A(\overline{\mathbb{Q}}) \times A(\overline{\mathbb{Q}}) \to \mathbb{R}$ induces an inner product on the $\mathbb{R}$-vector space $A(\overline{\mathbb{Q}}) \otimes \mathbb{R}$.*

Proof It suffices to show the positive definiteness of $\langle\,,\,\rangle_L$: if $x \in A(\overline{\mathbb{Q}}) \otimes \mathbb{R}$ satisfies $\langle x,x\rangle_L = 0$, then $x = 0$.

We write

$$x = a_1 x_1 + \cdots + a_m x_m$$

with $a_1, \ldots, a_m \in \mathbb{R}$ and $x_1, \ldots, x_m \in A(\overline{\mathbb{Q}})$. By passing from K to a suitable finite field extension of K, we may assume $x_1, \ldots, x_m \in A(K)$.

Let V be the finite dimensional $\mathbb{R}$-subspace of $A(\overline{\mathbb{Q}}) \otimes \mathbb{R}$ generated by $x_1, \ldots, x_m$, and we set $d = \dim V$. Rearranging $x_1, \ldots, x_m$ if necessary, we may assume that $\{x_1, \ldots, x_d\}$ is a basis for V. We define a subset M of V by $M := \{n_1 x_1 + \cdots + n_d x_d \mid n_1, \ldots, n_d \in \mathbb{Z}\}$.

Let $\gamma \in M$. By Theorem 3.32, 2, if $\langle\gamma,\gamma\rangle_L = 0$, then $\gamma = 0$. (Here, if we regard γ as an element of $A(\overline{\mathbb{Q}})$ instead of an element of $A(\overline{\mathbb{Q}}) \otimes \mathbb{R}$, then it follows that $\gamma \in A(\overline{\mathbb{Q}})_{\text{tor}}$.) Further, as $\frac{1}{2}\langle\gamma,\gamma\rangle_L = \hat{h}_L(\gamma)$ and L is an ample line bundle, $\{\gamma \in M \mid \langle\gamma,\gamma\rangle_L \leq 1\}$ is a finite set by Northcott's finiteness theorem (Theorem 3.14). Thus, $\epsilon := \inf\{\langle\gamma,\gamma\rangle_L \mid \gamma \in M, \gamma \neq 0\}$ is a positive number.

Let (p,q) denote the signature of the quadratic form $\langle\,,\,\rangle_L$ on V. We prove $p = d$. To derive a contradiction, suppose that $p < d$. We can choose a basis $\{u_1, \ldots, u_d\}$ for V such that

$$\langle v,v\rangle_L = \sum_{i=1}^{p} c_i^2 - \sum_{j=p+1}^{p+q} c_j^2,$$

for any $v = \sum_{i=1}^{d} c_i u_i \in V$ ($c_i \in \mathbb{R}$). We endow V with the inner product such that $\{u_1, \ldots, u_d\}$ is an orthonormal basis for V, and we endow V with the normalized Lebesgue measure vol such that the volume of the hypercube spanned by $\{u_1, \ldots, u_d\}$ is one. If we set

$$S := \left\{ v = \sum_{i=1}^{d} c_i u_i \in V \;\middle|\; \sum_{i=1}^{p} c_i^2 \leq \epsilon/2, \; \sum_{j=p+1}^{d} c_j^2 \leq R \right\},$$

for a sufficiently large R, then S is a centrally symmetric convex body in V and $\text{vol}(S)$ is sufficiently large. Hence, by Minkowski's convex body theorem (Theorem 2.9), there is a nonzero element $\gamma \in S \cap M$, and γ satisfies $\langle\gamma,\gamma\rangle_L \leq \epsilon/2$. This contradicts with the choice of ϵ. It follows that $p = d$, so $\langle\,,\,\rangle_L$ induces a positive definite quadratic form on V. $\square$

Let K be a number field. As we will see in Theorem 3.42, $A(K)$ is a finitely generated abelian group. Then Proposition 3.33 tells us that the Néron–Tate height pairing $\langle\ ,\ \rangle_L: A(K) \times A(K) \to \mathbb{R}$ induces an inner product on the finite dimensional $\mathbb{R}$-vector space $A(K) \otimes \mathbb{R}$.

3.8 Curves and Their Jacobians

In this section, we apply results in the previous section to certain principally polarized abelian varieties, called Jacobian varieties. In the first half of this section, which is classical in algebraic geometry, we briefly review the required constructions over the field $\mathbb{C}$ of complex numbers, and then over an arbitrary field. In the latter half, we study canonical heights on Jacobian varieties.

Over $\mathbb{C}$, the Jacobian variety of a connected compact Riemann surface (Figure 3.3) C is given as follows (for details, we refer to, e.g., [10, Chapter 2] or [4, Chapter 11]). Let ω_C be the canonical line bundle on C, i.e., the invertible sheaf of 1-forms on C, let $H_1(C, \mathbb{Z})$ be the group of homology classes of 1-cycles on C. Let g be the *genus* of C, i.e.,

$$g := \dim_{\mathbb{C}} H^0(C, \omega_C) = \frac{1}{2} \mathrm{rk}_{\mathbb{Z}} H_1(C, \mathbb{Z}),$$

which is also equal to the number of the handles of C. The well-known Riemann bilinear relation implies the following results (see [10, p. 307] or [4, p. 316]).

a. The pairing

$$H^0(C, \omega_C) \times H_1(C, \mathbb{Z}) \to \mathbb{C}, \quad (\omega, \gamma) \mapsto \int_{\gamma} \omega,$$

Figure 3.3 Bernhard Riemann.
Source: The Smithsonian Libraries.

defines an injection from $H_1(C,\mathbb{Z})$ to the dual space of $H^0(C,\omega_C)$. We define the *Jacobian variety* J of C to be the complex torus

$$H^0(C,\omega_C)^\vee / H_1(C,\mathbb{Z}).$$

In particular, we can easily see that the cotangent space of J at 0 is naturally isomorphic to $H^0(C,\omega_C)$.

b. The class in $H^2(J,\mathbb{Z}) = \mathrm{Hom}(\wedge^2 H_1(C,\mathbb{Z}) \to \mathbb{Z})$ defined by the intersection pairing

$$\wedge^2 H_1(C,\mathbb{Z}) \to \mathbb{Z}, \quad (\gamma,\gamma') \mapsto \gamma \cap \gamma',$$

is the cohomology class of an effective and ample divisor on J. We denote this divisor by Θ. Then Θ is uniquely determined by this class up to translation in J.

We fix a base point $p_0 \in C$. We call the map $j_{p_0} : C \to J$ sending a point $p \in C$ to the class of the homomorphism

$$\omega \mapsto \int_{p_0}^{p} \omega$$

the *Abel–Jacobi map* associated to p_0. The following results are known.

c. (Abel–Jacobi theorem) The natural extension

$$\mathrm{Div}^0(C) \to J, \quad \sum_{P \in C} n_P P \mapsto \sum_{P \in C} n_P j_{p_0}(P),$$

does not depend on the choice of p_0. It is surjective and the kernel is exactly the subgroup generated by the principal divisors on C (see [10, p. 235] or [4, p. 319]).

d. The divisor Θ in Assertion b above coincides with the prime divisor

$$\overbrace{j_{p_0}(C) + \cdots + j_{p_0}(C)}^{g - 1 \text{ times}}$$

up to translation (see [10, p. 338] or [4, p. 324]).[2]

In the book [31], A. Weil gave a purely algebraic construction of the Jacobian variety which works over an arbitrary field. For modern treatment of the construction, we refer to, e.g., [11, p. 323] or [24, Part III]. (However, the algebraic construction of the Jacobian variety remains complicated.)

---

[2] The divisor Θ is also known to coincide with the zero divisor of the classical Riemann theta function, up to translation.

Figure 3.4 Jules Henri Poincaré.
Source: The Smithsonian Libraries.

Let F be a field of characteristic zero and let $\overline{F}$ be an algebraic closure of F. Let C be a geometrically irreducible, smooth projective curve over F. We assume that C has an F-rational point. In the following, we briefly describe an algebraic construction of the Jacobian variety of C.

As in the case over $\mathbb{C}$, the *genus* of C is defined as

$$g := \dim_F H^0(C, \omega_C) = \dim_F H^1(C, \mathcal{O}_C),$$

where ω_C denotes the canonical line bundle on C.

Recall from N10 on p. 164 that the Picard group $\text{Pic}(C)$ is an abelian group consisting of all isomorphism classes of line bundles on C. For any integer $d \in \mathbb{Z}$, we define $\text{Pic}^d(C)$ to be the subset of $\text{Pic}(C)$ consisting of the classes of line bundles of degree d on C.

As in the case over $\mathbb{C}$, $\text{Pic}^d(C)$ is endowed with the structure of an algebraic variety over F, which we call the *Picard variety*[3] of C *of degree* d. The *Poincaré line bundle of degree* d (Figure 3.4) is a line bundle Q_d on $C \times \text{Pic}^d(C)$ such that $Q_d|_{C \times \{L\}} \cong L$ for any $L \in \text{Pic}^d(C)$. If $p \colon C \times \text{Pic}^d(C) \to \text{Pic}^d(C)$ denotes the second projection, then Q_d is uniquely determined modulo $p^*(\text{Pic}(\text{Pic}^d(C)))$.

Now we let $d = 0$. Then $\text{Pic}^0(C)$ is a geometrically irreducible, projective variety over F. Further, $\text{Pic}^0(C)$ is equipped with the group structure with binary operation given by the tensor $\otimes$. Thus, $\text{Pic}^0(C)$ is an abelian variety of dimension g over F. We set $J := \text{Pic}^0(C)$, and call J the *Jacobian variety* (Figure 3.5) of C.

In the following, we often identify a Cartier divisor D with the corresponding line bundle $\mathcal{O}(D)$.

---

[3] In this book, where no confusion is likely, we will use the same notation for the Picard variety $\text{Pic}^d(C)$ of C of degree d and its underlying set.

C. G. J. Jacobi.

Figure 3.5 Carl Jacobi.
Source: The Smithsonian Libraries.

Let Δ be the diagonal divisor of $C \times C$ and let $p_i \colon C \times C \to C$ be the i-th projection for $i = 1, 2$. Let a be a line bundle of degree one on C. We define the *Abel–Jacobi map* $j_a \colon C \to J$ associated to a by the map sending $x \in C(\overline{F})$ to $\mathcal{O}_C(x) \otimes a^{-1}$, which is often denoted by $x - a$ in the additive notation.

Theorem 3.34 (Injectivity of Abel–Jacobi maps) *If $g \geq 1$, then $j_a \colon C \to J$ is a closed immersion for any $a \in \mathrm{Pic}^1(C)$.*

Proof The assertion is well-known, so we will just sketch a proof. We may assume that $F = \overline{F}$. First, we show the injectivity of j_a. To derive a contradiction, we suppose that there are two distinct points $x, y \in C$ such that $j_a(x) = j_a(y)$. Then $x \sim y$ as divisors, i.e., there is a nonzero rational function f on C such that $y = x + (f)$. We set $L = \mathcal{O}_C(x)$, and we write s_1 and s_2 for the global sections of L corresponding to 1 and f, respectively. Since $\mathrm{div}(s_1) = x$ and $\mathrm{div}(s_2) = y$, we can define a morphism $\pi \colon C \to \mathbb{P}^1$ by $\zeta \mapsto (s_1(\zeta) : s_2(\zeta))$. Further, since $\pi^*(\mathcal{O}_{\mathbb{P}^1}(1)) \cong L$, we have $\deg(\pi) = 1$, i.e., π is an isomorphism. This contradicts the assumption that $g \geq 1$.

It remains to show that j_a induces surjections of the cotangent spaces. Let $x \in C$. The cotangent space of J at $j_a(x)$ is naturally identified with $H^0(C, \omega_C)$, and the induced homomorphism coincides with the evaluation map $H^0(C, \omega_C) \to \omega_C(x)$ $(\lambda \mapsto \lambda(x))$ via the identification. If $g \geq 1$, then ω_C is base point free, so the evaluation map is surjective. This completes the proof. $\square$

In the rest of this section, as in the case over $\mathbb{C}$, we introduce a *theta divisor* on the Jacobian variety J. We claim that, replacing F by a suitable finite extension field if necessary, there is $\theta \in \mathrm{Pic}^1(C)$ with $(2g - 2)\theta \sim \omega_C$. Indeed, we take any $\theta_0 \in C(\overline{F})$. We have seen that $J(\overline{F})$ is a divisible group (Corollary 3.25), so $\omega_C - (2g - 2)\theta_0 \sim (2g - 2)\theta_1$ for some $\theta_1 \in J(\overline{F})$. Replacing F with a suitable finite field extension of F, we may assume that $\theta_0, \theta_1 \in \mathrm{Pic}^1(C)$. Then $\theta := \theta_0 + \theta_1 \in \mathrm{Pic}^1(C)$ satisfies $(2g - 2)\theta \sim \omega_C$, and we obtain the claim.

To simplify the notation, we denote the Abel–Jacobi map j_θ by j. The theta divisor Θ on J (with respect to θ) is defined by

$$\Theta = \overbrace{j(C) + \cdots + j(C)}^{g - 1 \text{ times}}.$$

Note that Θ depends on the choice of θ. (To be precise, Θ depends on the choice of a half canonical divisor $\kappa = (g - 1)\theta$ with $2\kappa \sim \omega_C$.)

We will now show some properties of Θ.

Proposition 3.35 *1. $\mathcal{O}_J(\Theta)$ is an even line bundle on J.*
2. For any $a \in \mathrm{Pic}^1(C)$, $j_a^*(\Theta) \sim (g - 1)\theta + a$. In particular, $j^*(\Theta) \sim g\theta$.
3. Let $p_i : C \times C \to C$ be the i-th projection and let $q_i : J \times J \to J$ be the i-th projection. Let Δ be the diagonal divisor of $C \times C$. We set $P = (q_1 + q_2)^*(\Theta) - q_1^*(\Theta) - q_2^*(\Theta)$. Then

$$(j \times j)^*(P) \sim p_1^*(\theta) + p_2^*(\theta) - \Delta.$$

4. $\mathcal{O}_J(\Theta)$ is an ample line bundle.

To show Proposition 3.35, we would like to assume that $\overline{F} = F$. The next lemma is useful for this purpose.

Lemma 3.36 *Let V be a d-dimensional projective variety over F and let L and M be line bundles on V. Set $V_{\overline{F}} := V \times \mathrm{Spec}(\overline{F})$, $L_{\overline{F}} := L \otimes_F \overline{F}$, and $M_{\overline{F}} := M \otimes_F \overline{F}$. If[4] $L_{\overline{F}} \cong M_{\overline{F}}$ on $V_{\overline{F}}$, then $L \cong M$ on V.*

Proof Considering $L \otimes M^{-1}$ and $\mathcal{O}_V$ in place of L and M, respectively, we may assume $M = \mathcal{O}_V$. It follows from $L_{\overline{F}} \cong \mathcal{O}_{V_{\overline{F}}}$ that

$$\begin{cases} 0 \neq H^0(V_{\overline{F}}, L_{\overline{F}}) = H^0(V, L) \otimes_F \overline{F}, \\ 0 \neq H^0(V_{\overline{F}}, L_{\overline{F}}^{-1}) = H^0(V, L^{-1}) \otimes_F \overline{F}. \end{cases}$$

Thus, $H^0(V, L) \neq 0$ and $H^0(V, L^{-1}) \neq 0$. As in the proof of the seesaw theorem (Lemma 3.19), we have $L \cong \mathcal{O}_V$. $\qquad\square$

---

[4] Note that in Lemma 3.36 the given isomorphism $L_{\overline{F}} \cong M_{\overline{F}}$ need not coincide with the isomorphism $L_{\overline{F}} \cong M_{\overline{F}}$ induced from $L \cong M$.

Proof of Proposition 3.35   Note that Assertion 1 concerns the existence of an isomorphism $\mathcal{O}_J(\Theta) \cong [-1]_J^*(\mathcal{O}_J(\Theta))$ and that Assertions 2 and 3 concern the existence of isomorphisms of corresponding line bundles. Also, Assertion 4 holds over F if and only if it holds over $\overline{F}$. By Lemma 3.36, we may assume that F is algebraically closed.

1. It suffices to show that "$x \in \Theta \Longrightarrow -x \in \Theta$." If $x \in \Theta$, then by the definition of Θ, there are $g - 1$ points $P_1, \dots, P_{g-1} \in C$ such that

$$x \sim (P_1 - \theta) + \cdots + (P_{g-1} - \theta).$$

Since $\dim_F H^0(C, \omega_C) = g$,

$$H^0(C, \omega_C(-P_1 - \cdots - P_{g-1})) \neq 0.$$

Thus, there are $g - 1$ points $Q_1, \dots, Q_{g-1} \in C$ such that

$$(2g - 2)\theta \sim \omega_C \sim P_1 + \cdots + P_{g-1} + Q_1 + \cdots + Q_{g-1}.$$

It follows that

$$(Q_1 - \theta) + \cdots + (Q_{g-1} - \theta) = (Q_1 + \cdots + Q_{g-1}) - (g - 1)\theta$$
$$\sim (g - 1)\theta - (P_1 + \cdots + P_{g-1}) \sim -x,$$

so $-x \in \Theta$.

2. Note that $\{L \in \mathrm{Pic}^g(C) \mid \dim_F H^0(C, L) = 1\}$ is a Zariski open set of $\mathrm{Pic}^g(C)$ and the natural morphism $C^g \to \mathrm{Pic}^g(C)$ given by

$$(P_1, \dots, P_g) \mapsto \mathcal{O}_C(P_1 + \cdots + P_g)$$

is surjective, so if $a \in \mathrm{Pic}^1(C)$ is general, then

$$\dim_F H^0(C, \mathcal{O}_C(a + (g - 1)\theta)) = 1,$$

and the unique effective divisor belonging to $|a + (g-1)\theta|$ is the sum of distinct g points. Thus, there are distinct g points $P_1, \dots, P_g \in C$ that are unique up to reordering such that

$$P_1 + \cdots + P_g \sim a + (g - 1)\theta.$$

By Assertion 1 and the uniqueness of $P_1, \dots, P_g$ up to reordering, we have

$$
\begin{aligned}
j_a(x) \in \Theta &\Longleftrightarrow -j_a(x) \in \Theta \\
&\Longleftrightarrow a - x \sim (Q_1 - \theta) + \cdots + (Q_{g-1} - \theta) \quad (\exists\, Q_1, \dots, Q_{g-1} \in C) \\
&\Longleftrightarrow a + (g - 1)\theta \sim x + Q_1 + \cdots + Q_{g-1} \quad (\exists\, Q_1, \dots, Q_{g-1} \in C) \\
&\Longleftrightarrow P_1 + \cdots + P_g \sim x + Q_1 + \cdots + Q_{g-1} \quad (\exists\, Q_1, \dots, Q_{g-1} \in C) \\
&\Longleftrightarrow x \in \{P_1, \dots, P_g\}.
\end{aligned}
$$

Thus, $j_a^{-1}(\Theta) = \{P_1, \ldots, P_g\}$, and we have

$$j_a^*(\Theta) \sim P_1 + \cdots + P_g \sim a + (g-1)\theta.$$

Next, we consider the general case where $a \in \mathrm{Pic}^1(C)$ is arbitrary (not necessarily general). Let Q_g be a Poincaré line bundle of degree g on $C \times \mathrm{Pic}^g(C)$. Let $\alpha \colon C \times \mathrm{Pic}^1(C) \to C \times \mathrm{Pic}^g(C)$ be the morphism defined by $(x, a) \mapsto (x, (g-1)\theta + a)$ and let Q denote the pullback $\alpha^*(Q_g)$ of Q_g by α. Then, for any $a \in \mathrm{Pic}^1(C)$, we have $Q|_{C \times \{a\}} = (g-1)\theta + a$. Let $\Phi \colon C \times \mathrm{Pic}^1(C) \to J$ be the morphism defined by $\Phi(x, a) = x - a$. We set $N := \mathcal{O}_{C \times \mathrm{Pic}^1(C)}(\Phi^*(\Theta)) \otimes Q^{-1}$. We write $p \colon C \times \mathrm{Pic}^1(C) \to \mathrm{Pic}^1(C)$ for the second projection. A key observation is

$$N|_{p^{-1}(a)} = j_a^*(\Theta) - (g-1)\theta - a$$

for any $a \in \mathrm{Pic}^1(C)$. If $a \in \mathrm{Pic}^1(C)$ is general, then the above argument gives $\dim H^0(C, N|_{p^{-1}(a)}) = \dim H^0(C, \mathcal{O}_C) = 1 > 0$. Thus, by the upper semicontinuity of the dimension of the cohomology group, we have $\dim H^0(C, N|_{p^{-1}(a)}) > 0$ for any $a \in \mathrm{Pic}^1(C)$. On the other hand, since p is flat, $\deg(N|_{p^{-1}(a)}) = 0$ for all $a \in \mathrm{Pic}^1(C)$. Thus, $N|_{p^{-1}(a)} \cong \mathcal{O}_C$ for any $a \in \mathrm{Pic}^1(C)$.

3. We set $L = \mathcal{O}_{C \times C}(p_1^*(\theta) + p_2^*(\theta) - \Delta - (j \times j)^*(P))$. We claim that

$$L|_{\{x_1\} \times C} \cong \mathcal{O}_C \quad \text{and} \quad L|_{C \times \{x_2\}} \cong \mathcal{O}_C$$

for any $x_1, x_2 \in C$. Indeed, we define the morphism $f \colon C \to J \times J$ by $f(x) = (j(x_1), j(x))$. Then

$$(j \times j)^*(P)|_{\{x_1\} \times C} = f^*(P).$$

Since $j(x_1) + j(x) = x - (\theta - x_1 + \theta)$, Assertion 2 gives

$$\begin{aligned} f^*(P) &= j_{\theta - x_1 + \theta}^*(\Theta) - 0 - j^*(\Theta) \\ &= \theta - x_1 + \theta + (g-1)\theta - g\theta \\ &= \theta - x_1. \end{aligned}$$

On the other hand,

$$(p_1^*(\theta) + p_2^*(\theta) - \Delta)|_{\{x_1\} \times C} = \theta - x_1.$$

Thus, we get $L|_{\{x_1\} \times C} \cong \mathcal{O}_C$. We similarly have $L|_{C \times \{x_2\}} \cong \mathcal{O}_C$, and we obtain the claim.

Since $L|_{\{x_1\} \times C} \cong \mathcal{O}_C$, $p_{1*}(L)$ becomes a line bundle and $p_1^*(p_{1*}(L)) \to L$ is surjective. Thus, if we set $M = p_{1*}(L)$, then $L = p_1^*(M)$. On the other hand, it follows from $L|_{C\times\{x_2\}} \cong \mathcal{O}_C$ that $M \cong \mathcal{O}_C$. We obtain $L \cong \mathcal{O}_{C\times C}$. This completes the proof of Assertion 3.

4. For any $c \in J$, we define the morphism $T_c : J \to J$ by $T_c(x) :=$ $x+c$. Corollary 3.28 tells us that $\mathcal{O}_J(\Theta)$ is ample if and only if $\mathrm{Ker}(\lambda_{\mathcal{O}_J(\Theta)})$ is finite. Thus, it suffices to show that $\mathrm{Ker}(\lambda_{\mathcal{O}_J(\Theta)})$ is trivial, i.e., for any $c \in J$, $T_c^*(\Theta) \sim \Theta$ implies $c = 0$. It follows from $j^*(T_c^*(\Theta)) = j_{\theta-c}^*(\Theta)$ and Assertion 2 that

$$(g-1)\theta + (\theta - c) \sim j_{\theta-c}^*(\Theta) = j^*(T_c^*(\Theta)) \sim j^*(\Theta) \sim g\theta.$$

We obtain that $c \sim 0$ as divisors, so $c = 0$ as an element of J. $\square$

As an application of Proposition 3.35, we show some properties of heights. Let C be a geometrically irreducible, smooth projective curve over $\overline{\mathbb{Q}}$, and let θ be an element of $\mathrm{Pic}^1(C)$ with $(2g-2)\theta \sim \omega_C$. The theta divisor Θ (with respect to θ) is an even and ample divisor on J, so we have the Néron–Tate height pairing

$$\langle\,,\,\rangle_\Theta : J(\overline{\mathbb{Q}}) \times J(\overline{\mathbb{Q}}) \to \mathbb{R} \tag{3.13}$$

of $\mathcal{O}_J(\Theta)$ defined in Section 3.7. We show the following:

Corollary 3.37   *We keep the notation in Proposition 3.35.*

1. *We put $\Delta' = \Delta - p_1^*(\theta) - p_2^*(\theta)$ on $C \times C$. Then, for any $x, y \in C(\overline{\mathbb{Q}})$,*

$$\langle j(x), j(y)\rangle_\Theta = h_{-\Delta'}(x, y) + O(1).$$

2. *For any $x \in C(\overline{\mathbb{Q}})$,*

$$\langle j(x), j(x)\rangle_\Theta = 2gh_\theta(x) + O(1).$$

Proof   1. By Theorem 3.32,

$$\langle j(x), j(y)\rangle_\Theta = \hat{h}_P(j(x), j(y)),$$

where $P = (q_1+q_2)^*(\Theta) - q_1^*(\Theta) - q_2^*(\Theta)$. On the other hand, by Proposition 3.35, we have

$$\hat{h}_P(j(x), j(y)) = h_{(j\times j)^*(P)}(x, y) + O(1) = h_{-\Delta'}(x, y) + O(1).$$

2. By Proposition 3.35, we have

$$\langle j(x), j(x)\rangle_\Theta = 2\hat{h}_\Theta(j(x)) = 2h_{j^*(\Theta)}(x) + O(1)$$
$$= 2h_{g\theta}(x) + O(1) = 2gh_\theta(x) + O(1),$$

as required. $\square$

3.9 The Mordell–Weil Theorem

This section is devoted to proving the Mordell–Weil theorem (Figure 3.6), which asserts that, for any number field K and for any abelian variety A, the abelian group of the K-valued points $A(K)$ on A is finitely generated. The Mordell–Weil theorem is fundamental in arithmetic geometry, and it opens up truly various research directions. The Mordell–Weil theorem for elliptic curves is already very rich.

In this section, we first prove the Hermite–Minkowski theorem. Then from it we deduce the weak Mordell–Weil theorem. Finally, we combine the weak Mordell–Weil theorem and the Northcott finiteness theorem on heights to prove the Mordell–Weil theorem.

Theorem 3.38 (Hermite–Minkowski theorem)   *Let N be a positive integer. Then there are only finitely many number fields K such that $|D_{K/\mathbb{Q}}| \leq N$.*

Proof   Let K be a number field with $|D_{K/\mathbb{Q}}| \leq N$. By Minkowski's discriminant theorem (Theorem 2.13), there is a positive constant M depending only on N such that $[K : \mathbb{Q}] \leq M$. (For example, we can take $M := \log(4N)/\log(\pi)$ by Remark 2.15.)

We use the notation in Theorem 2.13. In particular, $n := [K : \mathbb{Q}]$, $\rho_1, \ldots, \rho_{r_1}$ are the real embeddings of K, and $\sigma_1, \bar{\sigma}_1, \ldots, \sigma_{r_2}, \bar{\sigma}_{r_2}$ are the embeddings of K into $\mathbb{C}$ that are not real. Let $V := K \otimes_{\mathbb{Q}} \mathbb{R}$ and let $f : V \to \mathbb{R}^n$ be the isomorphism of vector spaces defined by Equation (2.14). Further, we fix a constant C with $C > (2^3/\pi)^M \sqrt{|D_{K/\mathbb{Q}}|}$.

We may assume that $K \neq \mathbb{Q}$. First, we consider the case where $r_1 \geq 1$. This means that K has at least one real embedding, so we regard $K \subseteq \mathbb{R}$ via $\rho_1 : K \to \mathbb{R}$. We define the subset T of $\mathbb{R}^n$ by

Figure 3.6 André Weil.
Source: Archives of the Mathematisches Forschungsinstitut Oberwolfach.

$$\left\{ (a_1, \ldots, a_{r_1}, b_1, c_1, \ldots, b_{r_2}, c_{r_2}) \in \mathbb{R}^n \; \middle| \; \begin{array}{l} |a_1| < C, \; |a_i| < 1 \; (i = 2, \ldots, r_1), \\ \sqrt{2(b_j^2 + c_j^2)} < 1 \; (j = 1, \ldots, r_2) \end{array} \right\}.$$

Then $\mathrm{vol}(T) = C2^{r_1 - r_2} \pi^{r_2}$.

Let $S := f^{-1}(T) \subseteq V$, which is a centrally symmetric convex body in V. Note that we have endowed V with the inner product such that f is an isometry. Thus, $\mathrm{vol}(S) = \mathrm{vol}(T) = C2^{r_1 - r_2} \pi^{r_2}$. Since $\mathrm{vol}(O_K, \langle \, , \, \rangle_K) = \sqrt{|D_{K/\mathbb{Q}}|}$ by Lemma 2.12, we have

$$\mathrm{vol}(S) > (2^3/\pi)^{r_2} 2^{r_1 - r_2} \pi^{r_2} \sqrt{|D_{K/\mathbb{Q}}|} = 2^n \, \mathrm{vol}(O_K, \langle \, , \, \rangle_K)$$

by the assumption on C. We apply Minkowski's convex body theorem (Theorem 2.9) to find a nonzero element α in O_K such that $\alpha \in S$.

The condition $\alpha \in S$ implies that $|\alpha|_{\rho_1} < C$ and $|\alpha|_{\sigma} < 1$ for every $\sigma \in K(\mathbb{C})$ with $\sigma \neq \rho_1$. Thus, it follows from the condition $\alpha \in O_K \cap S$ and the product formula that $1 = \prod_{v \in M_K} |\alpha|_v \leq \prod_{\sigma \in K(\mathbb{C})} |\alpha|_\sigma < |\alpha|_{\rho_1}$.

Here, we claim the following:

Claim 2   If $\sigma \in K(\mathbb{C})$ satisfies $\sigma(\alpha) = \alpha$, then $\sigma = \rho_1$. Further, we have $K = \mathbb{Q}(\alpha)$.

Since we are regarding $K \subseteq \mathbb{R}$ via ρ_1, the condition $\sigma(\alpha) = \alpha$ implies $|\alpha|_\sigma = |\alpha|_{\rho_1}$. Since $|\cdot|_{\rho_1}$ is the only archimedean absolute value on K such that the absolute value of α is greater than one, we have $\sigma = \rho_1$. Thus, we obtain the first half of the claim.

An embedding $\sigma \in K(\mathbb{C})$ satisfies $\sigma(\alpha) = \alpha$ if and only if it is $\mathbb{Q}(\alpha)$-linear. Thus, there are $[K : \mathbb{Q}(\alpha)]$ such embeddings. By the first half of the claim, there is only one such embedding, so $[K : \mathbb{Q}(\alpha)] = 1$. Thus, $K = \mathbb{Q}(\alpha)$. We obtain Claim 2.

By Proposition 3.5, 5, the element α satisfies $h(\alpha) \leq \frac{1}{[K:\mathbb{Q}]} \log^+(C) \leq \log^+(C)$ and $[\mathbb{Q}(\alpha) : \mathbb{Q}] \leq M$, so there are only finitely many possibilities for such an element α by Northcott's finiteness theorem (Lemma 3.13). Since $K = \mathbb{Q}(\alpha)$ by Claim 2, it follows that there are only finitely many possibilities for such K. Since both C and M are constants depending only on N, we conclude that there are only finitely many number fields K having at least one real embedding and satisfying $|D_{K/\mathbb{Q}}| \leq N$.

Next, we consider the case where $r_1 = 0$. This means that K has no real embedding. In this case, $n = 2r_2$. We note that $\left| D_{K(\sqrt{-1})/\mathbb{Q}} \right|$ is bounded from above by a constant depending only on N. If we show that there are finitely many possibilities of $K(\sqrt{-1})$, then we see that there will be finitely many possibilities of K. Thus, we may replace K with $K(\sqrt{-1})$ and assume that K contains $\sqrt{-1}$. Put

$$
T' := \left\{ (b_1, c_1, \ldots, b_{r_2}, c_{r_2}) \in \mathbb{R}^n \;\middle|\; \begin{array}{l} \sqrt{2(b_1^2 + c_1^2)} < \sqrt{C}, \\[2mm] \sqrt{2(b_j^2 + c_j^2)} < 1 \; (j = 2, \ldots, r_2) \end{array} \right\}
$$

and set $S' := f^{-1}(T')$. By using Minkowski's convex body theorem (Theorem 2.9), we find a nonzero element α' in $O_K \cap S'$. We regard $K \subseteq \mathbb{C}$ via $\sigma_1 \colon K \to \mathbb{C}$. If $\sigma \in K(\mathbb{C})$ satisfies $\sigma(\alpha') = \alpha'$ and $\sigma(\sqrt{-1}) = \sqrt{-1}$, then $\sigma = \sigma_1$ by the same argument as above. Then we have $K = \mathbb{Q}(\alpha', \sqrt{-1})$ and we conclude that there are only finitely many such K. $\qquad\square$

Let K be a number field and let S be a finite subset of nonzero prime ideals of O_K. A field extension K' of K is said to be unramified outside S if any nonzero prime ideal P of O_K with $P \notin S$ is unramified (see Section 2.6).

Theorem 3.39 *Let K be a number field, let S be a finite subset of nonzero prime ideals of O_K, and let n be a positive integer. Then there are only finitely many field extensions K' of K such that $[K' : K] \le n$ and K' is unramified outside S.*

Proof Let $\pi \colon \mathrm{Spec}(O_K) \to \mathrm{Spec}(\mathbb{Z})$ be the natural map. We put $[K : \mathbb{Q}] = m$. It suffices to show that there are only finitely many field extensions K' of $\mathbb{Q}$ such that $[K' : \mathbb{Q}] \le nm$ and K' is unramified outside $\pi(S)$. Thus, we may assume that $K = \mathbb{Q}$ (so $m = 1$). Let K' be a field extension of $\mathbb{Q}$ such that $[K' : \mathbb{Q}] \le n$ and unramified outside S. We put $N = \prod_{p \in S} p^{n-1+n\log_p(n)}$. Then Theorem 2.18 tells us that $|D_{K'/\mathbb{Q}}| \le N$. According to the Hermite–Minkowski theorem (Theorem 3.38), there are only finitely many such K'. $\qquad\square$

Let D be a Dedekind domain, and we set $S = \mathrm{Spec}(D)$. In the notation at the beginning of Section 3.6, we replace a base field F with D. Then we have the notion of a *group scheme* $\mathcal{G}$ over S with three morphisms $m_{\mathcal{G}}, i_{\mathcal{G}}$, and $e_{\mathcal{G}}$. A group scheme $\mathcal{A}$ over S is called an *abelian scheme* over S if $\mathcal{A}$ is smooth and projective over S whose all geometric fibers are connected. Then each geometric fiber of $\mathcal{A}$ is an abelian variety with the three morphisms given by the restrictions of $m_{\mathcal{A}}, i_{\mathcal{A}}$, and $e_{\mathcal{A}}$ to the geometric fiber.

Lemma 3.40 *Let K be a number field and let A be an abelian variety over K. There then are a nonempty open subscheme $U \subseteq \mathrm{Spec}(O_K)$ and an abelian scheme $\mathcal{A}$ over U such that A is the restriction of $\mathcal{A}$ to the generic fiber.*

Proof Since A, m_A, i_A, and e_A are defined by finitely many polynomials with coefficients in K in suitable projective spaces, we can take $f \in O_K \setminus \{0\}$ such that all these polynomials multiplied by f have coefficients in O_K. We set $U := \mathrm{Spec}(O_K[1/f])$. By shrinking U if necessary, there are a smooth and projective scheme $\mathcal{A}$ over U that extends A and morphisms $m_{\mathcal{A}}, i_{\mathcal{A}}$, and $e_{\mathcal{A}}$ over U that extend m_A, i_A, and e_A, respectively, such that the diagrams in Section 3.6 commute. Then $\mathcal{A}$ becomes an abelian scheme over U with the morphisms $m_{\mathcal{A}}, i_{\mathcal{A}}$, and $e_{\mathcal{A}}$. $\square$

Theorem 3.41 (weak Mordell–Weil theorem)   *Let K be a number field, let $n \geq 2$ be an integer, and let A be an abelian variety over K. If all n-torsion points of A are defined over K, then $A(K)/nA(K)$ is a finite group.*

Proof By Lemma 3.40, A extends to an abelian scheme $\mathcal{A}$ over a nonempty open subscheme $U \subseteq \mathrm{Spec}(O_K)$. By shrinking U if necessary, we may assume that n is invertible on U. Consider the morphism $[n]_{\mathcal{A}} : \mathcal{A} \to \mathcal{A}$ defined by the multiplication by n. We claim that $[n]_{\mathcal{A}}$ is finite and étale. Indeed, Corollary 3.25 tells us that $[n]_{\mathcal{A}}$ is finite and flat. To show that $[n]_{\mathcal{A}}$ is étale, it suffices to show that $[n]_{\mathcal{A}}$ is étale on each fiber $\mathcal{A}_s$ over $s \in U$. Since $[n]_{\mathcal{A}_s}$ has degree n^{2g} by Corollary 3.25, the extension of the rational function fields induced by $[n]_{\mathcal{A}_s}$ is separable. Thus, there is a nonempty open subset of $\mathcal{A}_s$ on which $[n]_{\mathcal{A}_s}$ is étale. Considering translations, we see that $[n]_{\mathcal{A}_s}$ is étale on $\mathcal{A}_s$. Thus, we have shown the claim. Next, we show the following claim:

Claim 3   For any $P \in A(K)$, there are a finite field extension K'/K and a point $Q \in A(K')$ with the following properties: $nQ = P$, $[K' : K] \leq n^{2g}$, and $O_{K'}$ is étale over U.

Indeed, since $\mathcal{A}$ is projective over U, by the valuative criterion for properness, P extends to a morphism $\widetilde{P} : U \to \mathcal{A}$. Let $Z \to \mathcal{A}$ be the fiber product of $\widetilde{P}$ and $[n]_{\mathcal{A}}$. Since the morphism $Z \to U$ is finite and étale, Z is a regular affine scheme. Let Z' be a connected component of Z and let K' denote the total quotient field of Z'. Then Z' is étale over U and $[K' : K] \leq \deg([n]_{\mathcal{A}})$. Further, the morphism $Z' \hookrightarrow Z \to \mathcal{A}$ defines a point $Q \in A(K')$ such that $[n](Q) = P$. This completes the proof of Claim 3.

In Claim 3, $O_{K'}$ is, in particular, unramified over U. Thus, by Theorem 3.39, there are only finitely many such K'. It follows that there is a finite Galois extension L of K that contains all such K'. Let G denote the Galois group of L/K. We denote the group of n-torsion points of A by A_n. By the hypothesis of Theorem 3.41, we have $A_n \subseteq A(K)$.

Now we will construct a group homomorphism

$$\varphi : A(K) \to \mathrm{Hom}(G, A_n).$$

For $P \in A(K)$, we take $Q \in A(L)$ such that $nQ = P$. (The existence of such Q is guaranteed by Claim 3.) We take $g \in G$ and consider $g(Q) - Q$. First, since

$$n(g(Q) - Q) = g(nQ) - nQ = g(P) - P = P - P = 0,$$

we observe that $g(Q) - Q \in A_n$. We claim that $g(Q) - Q$ does not depend on the choice of Q with $nQ = P$. Indeed , let Q' be a point in $A(L)$ with $nQ' = P$. Then, since $n(Q' - Q) = P - P = 0$, we have $Q' - Q \in A_n \subseteq A(K)$. Thus,

$$(g(Q') - Q') - (g(Q) - Q) = g(Q' - Q) - (Q' - Q) = (Q' - Q) - (Q' - Q) = 0.$$

It follows that, if we define

$$\varphi \colon A(K) \to \mathrm{Map}(G, A_n)$$

by $\varphi(P)(g) = g(Q) - Q$, then φ is well-defined. For each $P \in A(K)$, $\varphi(P) \colon G \to A_n$ is in fact a group homomorphism, because

$$\varphi(P)(gg') = (gg')(Q) - Q = (g(Q) - Q) + g(g'(Q) - Q)$$
$$= (g(Q) - Q) + (g'(Q) - Q) = \varphi(P)(g) + \varphi(P)(g').$$

Thus, we obtain the required map $\varphi \colon A(K) \to \mathrm{Hom}(G, A_n)$. Further, φ is a group homomorphism. Indeed, for any $P_1, P_2 \in A(K)$, we choose $Q_1, Q_2 \in A(L)$ such that $nQ_1 = P_1$ and $nQ_2 = P_2$, respectively. Since $n(Q_1 + Q_2) = P_1 + P_2$,

$$\varphi(P_1 + P_2)(g) = g(Q_1 + Q_2) - (Q_1 + Q_2)$$
$$= (g(Q_1) - Q_1) + (g(Q_2) - Q_2)$$
$$= \varphi(P_1)(g) + \varphi(P_2)(g),$$

which implies that $\varphi(P_1 + P_2) = \varphi(P_1) + \varphi(P_2)$.

Now we consider the kernel of φ. Since

$$\varphi(P) = 0 \iff g(Q) - Q = 0 \ (\forall g \in G)$$
$$\iff Q \in \text{(the } G\text{-invariant subgroup of } A(L)) = A(K)$$
$$\iff P \in nA(K),$$

we have $\mathrm{Ker}(\varphi) = nA(K)$. Namely, $A(K)/nA(K)$ is isomorphic to a subgroup of $\mathrm{Hom}(G, A_n)$. Since $\mathrm{Hom}(G, A_n)$ is a finite group, we conclude that $A(K)/nA(K)$ is a finite group. $\qquad \square$

Theorem 3.42 (Mordell–Weil theorem) *Let K be a number field and let A be an abelian variety over K. Then $A(K)$ is a finitely generated abelian group.*

Proof Replacing K with a finite extension field if necessary, we may assume that all the 2-torsion points of A are contained in $A(K)$. We fix an even and ample line bundle L on A, and let $\langle \, , \, \rangle_L \colon A(K) \times A(K) \to \mathbb{R}$ be the Néron–Tate height pairing associated to L (see Section 3.7). We set $|x| = \sqrt{\langle x, x \rangle_L}$.

Suppose first that $|x| = 0$ for any $x \in A(K)$. Then $A(K) = A(K)_{\mathrm{tor}}$. We obtain by Northcott's finiteness theorem (Theorem 3.14) that $A(K)$ is a finite group in this case.

Suppose that $|x| > 0$ for some $x \in A(K)$. By the weak Mordell–Weil theorem (Theorem 3.41), $A(K)/2A(K)$ is a finite group. We take a complete set of representatives $\{x_i\}_{1 \le i \le a} \subseteq A(K)$ for $A(K)/2A(K)$ such that $|x_i| > 0$ for some i, and we set $C = \max_{1 \le i \le a}\{|x_i|\}$.

For each positive integer n, let B_n be the subgroup of $A(K)$ generated by $\{x \in A(K) \mid |x| \le nC\}$. Since $C > 0$, we have $A(K) = \bigcup_n B_n$ and

$$B_1 \subseteq B_2 \subseteq \cdots \subseteq B_n \subseteq \cdots .$$

We claim that, if $n \ge 3$, then $B_{n-1} = B_n$. Indeed, we take any $x \in A(K)$ with $|x| \le nC$. There then exist $y \in A(K)$ and x_i such that $x = 2y + x_i$. Since

$$|y| = \frac{|x - x_i|}{2} \le \frac{|x| + |x_i|}{2} \le \frac{n+1}{2} C$$

and $n \ge 3$, we have $|y| \le (n-1)C$. Thus, x is contained in the subgroup generated by y and x_i, so we have $x \in B_{n-1}$, which in turn implies that $B_{n-1} = B_n$. It follows that $A(K) = B_2$. Since $\{x \in A(K) \mid |x| \le 2C\}$ is finite by Northcott's finiteness theorem (Theorem 3.14), we obtain the assertion. $\qquad\square$

4

Preliminaries for the Proof
of Faltings's Theorem

This chapter is devoted to some fundamental results of Diophantine geometry such as Siegel's lemma (Lemma 4.1 and Proposition 4.3) and Roth's lemma (Theorem 4.20). We will use them in Chapter 5. Because our purpose is to give a proof of Faltings's theorem in not too many pages, we touch on only the essential results of Diophantine geometry. For more results of Diophantine geometry, see [6, 12, 13].

4.1 Siegel's Lemma

In this section, we prove Siegel's lemma (Figure 4.1). We will use it to find a small section of a line bundle in Chapter 5. First, we give a version for $\mathbb{Z}$.

Lemma 4.1 (Siegel's lemma for $\mathbb{Z}$)   *Let $A = (a_{ij})_{\substack{1 \le i \le m \\ 1 \le j \le n}}$ be an $m \times n$ matrix with entries in $\mathbb{Z}$, and let r be the rank of A. If $n > r \ge 1$, then there is a vector $x = {}^t(x_1, \ldots, x_n) \in \mathbb{Z}^n$ such that $x \ne 0$, $Ax = 0$ and*

$$\max_i \{|x_i|\} \le \left(n \max_{i,j} \{|a_{ij}|\} \right)^{\frac{r}{n-r}}.$$

Figure 4.1 Carl Siegel.
Source: Archives of the Mathematisches Forschungsinstitut Oberwolfach.

Proof First, we assume that $m = r$. Let $Q := \max_{ij}\{|a_{ij}|\}$ let H be a positive integer satisfying $H \leq (nQ)^{\frac{m}{n-m}} < H + 1$. For $x = {}^t(x_1, \ldots, x_n) \in \mathbb{Z}^n$ with $0 \leq x_j \leq H$ $(j = 1, \ldots, n)$, we set $y = {}^t(y_1, \ldots, y_m) = Ax$. If n_i is the number of negative integers in $\{a_{i1}, \ldots, a_{in}\}$, then

$$-n_i QH \leq y_i = \sum_{j=1}^{n} a_{ij}x_j \leq (n - n_i)QH.$$

Note that

$$(H + 1)^n = (H + 1)^m(H + 1)^{n-m} > (H + 1)^m \left((nQ)^{\frac{m}{n-m}}\right)^{n-m}$$
$$= (H + 1)^m(nQ)^m \geq (nQH + 1)^m.$$

Since

$$\#\left\{(x_1, \ldots, x_n) \in \mathbb{Z}^n \mid 0 \leq x_j \leq H \ (j = 1, \ldots, n)\right\} = (H + 1)^n \quad \text{and}$$
$$\#\left\{(y_1, \ldots, y_m) \in \mathbb{Z}^m \mid -n_i QH \leq y_i \leq (n - n_i)QH \ (i = 1, \ldots, n)\right\}$$
$$= (nQH + 1)^m,$$

Dirichlet's box principle[1] (Figure 4.2) implies that there are

$$x', x'' \in \left\{(x_1, \ldots, x_n) \in \mathbb{Z}^n \mid 0 \leq x_j \leq H \ (j = 1, \ldots, n)\right\}$$

such that $x' \neq x''$ and $Ax' = Ax''$. We set $x = x' - x''$. Then x is a desired solution.

Next, we consider the general case. Renumbering the rows of A, we may assume that the first r rows of A are linearly independent over $\mathbb{Q}$. Let A' be the matrix consisting of the first r rows of A. Then, since any remained row of A can be written as a linear combination of the rows of A', for $x \in \mathbb{Z}^n$, "$Ax = 0$" is equivalent to "$A'x = 0$." Thus, applying the above assertion to A', we obtain the lemma.   □

Let K be a number field with $d = [K : \mathbb{Q}]$. As defined in (2.7), let $K(\mathbb{C})$ be the set of all embeddings of K into $\mathbb{C}$ as fields. For $x \in K$, we set

$$\|x\|_K = \max_{\sigma \in K(\mathbb{C})} \{|\sigma(x)|\}. \tag{4.1}$$

---

[1] It is also called the pigeonhole principle. Let X, Y be sets and let $f : X \to Y$ be a map. The principle asserts that if Y is a finite set, then there is $y \in Y$ such that $\#(f^{-1}(y)) \geq \#(X)/\#(Y)$. If X is also a finite set with $\#(X) > \#(Y)$, then the principle asserts that $\#(f^{-1}(y)) \geq 2$. IF X is an infinite set, the principle asserts that $f^{-1}(y)$ is an infinite set.

Figure 4.2 Lejeune Dirichlet.
Source: Getty Images / Hulton Archive / Stringer.

Let O_K be the ring of integers in K and $\{\omega_1, \ldots, \omega_d\}$ be a free basis of O_K over $\mathbb{Z}$. For $x = b_1\omega_1 + \cdots + b_d\omega_d \in K$ ($b_i \in \mathbb{Q}$), we set

$$\|x\|_0 = \max_i\{|b_i|\}. \tag{4.2}$$

Lemma 4.2 *Both $\| \ \|_K$ and $\| \ \|_0$ extend to the norms of the d-dimensional vector space $K \otimes_\mathbb{Q} \mathbb{R}$ over $\mathbb{R}$. Further, there are positive constants M_1 and M_2, depending on K and the choice of a free basis of O_K over $\mathbb{Z}$, such that $M_1\|x\|_K \le \|x\|_0 \le M_2\|x\|_K$ for all $x \in K \otimes_\mathbb{Q} \mathbb{R}$.*

Proof Note that

$$\|b_1\omega_1 + \cdots + b_l\omega_d\|_K = \max_{\sigma \in K(\mathbb{C})} \{|b_1\sigma(\omega_1) + \cdots + b_d\sigma(\omega_d)|\}$$

for $b_1, \ldots, b_d \in \mathbb{Q}$. By the same equation for $b_1, \ldots, b_d \in \mathbb{R}$, we define $\|\cdot\|_K$ on $K \otimes_\mathbb{Q} \mathbb{R}$. Then the extension does not depend on the choice of the free basis $\{\omega_1, \ldots, \omega_d\}$.

We need to show that $\|\cdot\|_K$ gives a norm on $K \otimes_\mathbb{Q} \mathbb{R}$. It suffices to show that $\|\cdot\|_K$ is positive definite. For $x \in K$, we set

$$\|x\| = \sqrt{\sum_{\sigma \in K(\mathbb{C})} |\sigma(x)|^2}.$$

As shown in Proposition 2.11, $\|\cdot\|$ extends to the norm of the vector space $K \otimes_\mathbb{Q} \mathbb{R}$ over $\mathbb{R}$. Further, the inequality

$$\|\cdot\| \le [K : \mathbb{Q}]^{1/2}\|\cdot\|_K$$

holds on K, and thus, on $K \otimes_\mathbb{Q} \mathbb{R}$. It follows that, if $x \in K \otimes_\mathbb{Q} \mathbb{R}$ satisfies $\|x\|_K = 0$, then $x = 0$. We have shown that $\|\cdot\|_K$ is a norm on $K \otimes_\mathbb{Q} \mathbb{R}$.

We extend $\|\cdot\|_0$ on K to $K \otimes_\mathbb{Q} \mathbb{R}$ by defining

$$\|x\|_0 = \max_i\{|b_i|\} \quad (x = b_1\omega_1 + \cdots + b_d\omega_d \in K \otimes_\mathbb{Q} \mathbb{R}, \ b_1, \ldots, b_d \in \mathbb{R}).$$

Then $\|\cdot\|_0$ gives a norm on the vector space $K \otimes_{\mathbb{Q}} \mathbb{R}$ over $\mathbb{R}$. Note that $\|\cdot\|_0$ depends on the choice of a free basis of O_K over $\mathbb{Z}$.

Identifying $K \otimes_{\mathbb{Q}} \mathbb{R} \ni \sum b_i \omega_i$ with $(b_i) \in \mathbb{R}^n$, we endow $K \otimes_{\mathbb{Q}} \mathbb{R}$ with the Euclidean topology from $\mathbb{R}^n$. Since $B := \{x \in K \otimes_{\mathbb{Q}} \mathbb{R} \mid \|x\|_K = 1\}$ is compact and $\|\cdot\|_0$ is a positive continuous function on B, there are positive numbers M_1 and M_2 such that

$$M_1 \le \|x\|_0 \le M_2$$

for all $x \in B$, from which the last assertion follows. $\square$

Proposition 4.3 (Siegel's lemma for algebraic integers)   *Let K be a number field and O_K be the ring of integers in K. Let $A = (a_{ij})_{\substack{1 \le i \le m \\ 1 \le j \le n}}$ be an $m \times n$ matrix with entries in O_K, and let r be the rank of A. If $n > r \ge 1$, then there is a vector $x = {}^t(x_1, \ldots, x_n) \in O_K^n$ such that $x \ne 0$, $Ax = 0$, and*

$$\max_i \{\|x_i\|_K\} \le c_1 \left(c_2 n \max_{i,j} \{\|a_{ij}\|_K\} \right)^{\frac{r}{n-r}},$$

where c_1 and c_2 are positive constants depending on K and the choice of a free basis of O_K over $\mathbb{Z}$, but not on A or x.

Proof   Let $d = [K : \mathbb{Q}]$, let $\{\omega_1, \ldots, \omega_d\}$ be a free basis of O_K over $\mathbb{Z}$, and let $\|\cdot\|_0$ be the norm given as (4.2).

For a vector $v = (v_i)$ and a matrix $A = (a_{ij})$ with entries in K, we set

$$\|v\|_K = \max_i \{\|v_i\|_K\} \quad \text{and} \quad \|A\|_K = \max_{i,j} \{\|a_{ij}\|_K\}.$$

Similarly, we define $\|v\|_0$ and $\|A\|_0$ using (4.2). With the constants M_1 and M_2 in Lemma 4.2, we have

$$M_1 \|v\|_K \le \|v\|_0 \le M_2 \|v\|_K, \quad \text{and} \quad M_1 \|A\|_K \le \|A\|_0 \le M_2 \|A\|_K.$$

We set

$$\omega_i \cdot \omega_j = \sum_{l=1}^{d} \alpha_{ijl} \omega_l \quad (\alpha_{ijl} \in \mathbb{Z}).$$

Let $\{e_j\}$ and $\{e_i'\}$ be the standard free bases of O_K^n and O_K^m as O_K-modules, respectively. Then $\{\omega_l e_j\}$ and $\{\omega_l e_i'\}$ give free bases over $\mathbb{Z}$. Let $\tilde{A}$ be the matrix representation of the map

$$O_K^n \to O_K^m \quad (x \mapsto Ax)$$

with respect to the free bases $\{\omega_t e_j\}$ and $\{\omega_l e_i'\}$ over $\mathbb{Z}$. Then $\widetilde{A}$ is an $md \times nd$ matrix with entries in $\mathbb{Z}$. Let Q be the maximum among the absolute values of entries of $\widetilde{A}$. Then, by Lemma 4.1, there is an $x \in O_K^n$ such that $x \neq 0$, $Ax = 0$ and

$$\|x\|_0 \leq (ndQ)^{\frac{dr}{dn-dr}} = (ndQ)^{\frac{r}{n-r}}.$$

We take $a_{ijs} \in \mathbb{Z}$ such that $a_{ij} = \sum_{s=1}^d a_{ijs}\omega_s$. Then $\|A\|_0 = \max\limits_{i,j,s}\{|a_{ijs}|\}$. Further, since the i-th entry of $A(\omega_t e_j)$ is

$$a_{ij}\omega_t = \left(\sum_{s=1}^d a_{ijs}\omega_s\right)\omega_t = \sum_{s=1}^d a_{ijs}\sum_{l=1}^d \alpha_{stl}\omega_l = \sum_{l=1}^d\sum_{s=1}^d (a_{ijs}\alpha_{stl})\omega_l,$$

we have $\widetilde{A} = \left(\sum_{s=1}^d (a_{ijs}\alpha_{stl})\right)_{(ij,tl)}$. Putting $\alpha = \max_{s,t,l}\{|\alpha_{stl}|\}$, we get $Q \leq d\|A\|_0\alpha$, so that

$$\|x\|_0 \leq (ndQ)^{\frac{r}{n-r}} \leq (nd^2\alpha\|A\|_0)^{\frac{r}{n-r}}.$$

Since $M_1\|x\|_K \leq \|x\|_0$ and $\|A\|_0 \leq M_2\|A\|_K$, we obtain the required estimate by setting $c_1 = 1/M_1$ and $c_2 = d^2\alpha M_2$. $\qquad\square$

4.2 Inequalities on Lengths and Heights of Polynomials

In this section, we define the Mahler measure of a polynomial and show some properties of heights of polynomials such as Gelfond's inequality.

Let K be a number field. For $I = (i_1, \ldots, i_m) \in \mathbb{Z}_{\geq 0}^m$, we write X^I for the monomial $X_1^{i_1}\cdots X_m^{i_m}$. Let $F = \sum_{I \in \mathbb{Z}_{\geq 0}^m} a_I X^I$ be an element of $K[X_1, \ldots, X_m]$. The *length* of F with respect to $v \in M_K$ is defined by $|F|_v = \max_{I \in \mathbb{Z}_{\geq 0}^m}\{|a_I|_v\}$. Note that $|\cdot|_v$ gives a norm, i.e., for any $F, G \in K[X_1, \ldots, X_m]$ and $a \in K$,

$$\begin{cases} |F + G|_v \leq |F|_v + |G|_v & \text{(if } v \text{ is archimedean)}, \\ |F + G|_v \leq \max\{|F|_v, |G|_v\} & \text{(if } v \text{ is nonarchimedean)}, \\ |aF|_v = |a|_v|F|_v, \\ |F|_v = 0 \iff F = 0. \end{cases} \qquad (4.3)$$

In the following, for $F, G \in K[X_1, \ldots, X_m]$, we compare $|FG|_v$ with $|F|_v|G|_v$ for each $v \in M_K$. As an application, we compare the height of FG and the sum of the heights of F and G (see Proposition 4.8).

Figure 4.3 Carl Friedrich Gauss.
Source: The Smithsonian Libraries.

4.2.1 Nonarchimedean Case

Suppose that v is nonarchimedean. Then the following Gauss's lemma (Figure 4.3) holds.

Proposition 4.4 (Gauss's lemma)   *For any $F, G \in K[X_1, \ldots, X_m]$, we have* $|FG|_v = |F|_v |G|_v$.

Proof   Let O_K be the ring of integers in K. Let P be the prime ideal corresponding to the absolute value $|\cdot|_v$. Let $F = \sum_I a_I X^I$, $G = \sum_I b_I X^I$ and $FG = \sum_I c_I X^I$. Then the assertion $|FG|_v = |F|_v |G|_v$ is equivalent to

$$\min_I \{\operatorname{ord}_P(a_I)\} + \min_I \{\operatorname{ord}_P(b_I)\} = \min_I \{\operatorname{ord}_P(c_I)\}.$$

We set

$$\begin{cases} a'_I = t^{-a} a_I, & b'_I = t^{-b} b_I, & c'_I = t^{-a-b} c_I, \\ F' = \sum_I a'_I X^I, & G' = \sum_I b'_I X^I, & H' = \sum_I c'_I X^I, \end{cases}$$

where t is a generator of $P(O_K)_P$,

$$a = \min_I \{\operatorname{ord}_P(a_I)\} \quad \text{and} \quad b = \min_I \{\operatorname{ord}_P(b_I)\}.$$

Then $H' = F'G'$ and $F', G', H' \in (O_K)_P[X_1, \ldots, X_m]$. Note that

$$\min_I \{\operatorname{ord}_P(a'_I)\} = 0 \quad \text{and} \quad \min_I \{\operatorname{ord}_P(b'_I)\} = 0.$$

Let J be the ideal of $(O_K)_P$ generated by $\{c'_I\}_I$. We claim that $J = (O_K)_P$. If not, $J \subseteq P(O_K)_P$, so if we denote the canonical image of F' and G' in $((O_K)_P/P(O_K)_P)[X_1, \ldots, X_m]$ by $\bar{F}'$ and $\bar{G}'$, respectively, then $\bar{F}'\bar{G}' = 0$. Since $((O_K)_P/P(O_K)_P)[X_1, \ldots, X_m]$ is an integral domain, we have $\bar{F}' = 0$

or $\bar{G}' = 0$, which means that $\min_I\{\mathrm{ord}_P(a'_I)\} > 0$ or $\min_I\{\mathrm{ord}_P(b'_I)\} > 0$. This is a contradiction. Thus, $J = (O_K)_P$, i.e., $\min_I\{\mathrm{ord}_P(c'_I)\} = 0$. It follows that $\min_I\{\mathrm{ord}_P(c_I)\} = a + b$. $\qquad\square$

4.2.2 Archimedean Case

Suppose that v is archimedean. We define $|F|_\infty$ to be

$$|F|_\infty = \max_{I \in \mathbb{Z}_{\geq 0}^m} \{|a_I|\}$$

for $F \in \sum_{I \in \mathbb{Z}_{\geq 0}^m} a_I X^I \in \mathbb{C}[X_1, \ldots, X_m]$, where $|a_I|$ is the usual absolute value of $a_I \in \mathbb{C}$. Clearly $|\cdot|_\infty$ is a $\mathbb{C}$-norm, i.e., for any $F, G \in \mathbb{C}[X_1, \ldots, X_m]$ and $a \in \mathbb{C}$,

$$\begin{cases} |F + G|_\infty \leq |F|_\infty + |G|_\infty, \\ |aF|_\infty = |a||F|_\infty, \\ |F|_\infty = 0 \iff F = 0. \end{cases} \tag{4.4}$$

We are going to compare $|F|_\infty |G|_\infty$ and $|FG|_\infty$ for $F, G \in \mathbb{C}[X_1, \ldots, X_m]$. For this purpose, we introduce the following quantities:

$$\begin{cases} M(F) = \exp\left(\int_0^1 \cdots \int_0^1 \log(|F(e^{2\pi\sqrt{-1}t_1}, \ldots, e^{2\pi\sqrt{-1}t_m})|) dt_1 \cdots dt_m \right), \\ L_2(F) = \left(\sum_{I \in \mathbb{Z}_{\geq 0}^m} |a_I|^2 \right)^{1/2}. \end{cases}$$

The invariant $M(F)$ is called the *Mahler measure* of F. The Mahler measure is multiplicative:

$$M(FG) = M(F)M(G) \quad (\forall\, F, G \in \mathbb{C}[X_1, \ldots, X_m]).$$

Indeed,

$$\log M(FG) = \int_0^1 \cdots \int_0^1 \log(|(FG)(e^{2\pi\sqrt{-1}t_1}, \ldots, e^{2\pi\sqrt{-1}t_m})|) dt_1 \cdots dt_m$$

$$= \int_0^1 \cdots \int_0^1 \log(|F(e^{2\pi\sqrt{-1}t_1}, \ldots, e^{2\pi\sqrt{-1}t_m})|) dt_1 \cdots dt_m$$

$$+ \int_0^1 \cdots \int_0^1 \log(|G(e^{2\pi\sqrt{-1}t_1}, \ldots, e^{2\pi\sqrt{-1}t_m})|) dt_1 \cdots dt_m$$

$$= \log M(F) + \log M(G),$$

as required.

In the following, the degree of F with respect to the variable X_i is denoted by $\deg_i(F)$ (see N5 on p. 163).

Proposition 4.5   *For nonzero polynomials $F, G \in \mathbb{C}[X_1, \ldots, X_m]$, we have the following:*

1. $|F|_\infty \leq L_2(F) \leq (\deg_1(F) + 1)^{1/2} \cdots (\deg_m(F) + 1)^{1/2} |F|_\infty$.
2. $M(F) \leq L_2(F)$.
3. $|F|_\infty \leq 2^{\deg_1(F) + \cdots + \deg_m(F)} M(F)$.
4. $|FG|_\infty \leq |F|_\infty |G|_\infty \prod_{i=1}^{m} (1 + \min\{\deg_i(F), \deg_i(G)\})$.
5. $|F|_\infty |G|_\infty \leq \exp(m(\deg(F) + \deg(G))) |FG|_\infty$.

Proof   1. The number of the monomials in F is at most

$$(\deg_1(F) + 1) \cdots (\deg_m(F) + 1).$$

Thus, the assertion follows.

2. For $n \in \mathbb{Z}$,

$$\int_0^1 e^{2n\pi\sqrt{-1}t} dt = \begin{cases} 0 & (\text{if } n \neq 0), \\ 1 & (\text{if } n = 0), \end{cases}$$

so direct calculations give

$$L_2(F) = \left(\int_0^1 \cdots \int_0^1 |F(e^{2\pi\sqrt{-1}t_1}, \ldots, e^{2\pi\sqrt{-1}t_m})|^2 dt_1 \cdots dt_m \right)^{1/2}.$$

Thus, it suffices to show that, for any C^2-class function φ on $\mathbb{R}$ with $\varphi'' \geq 0$ and any continuous function u on $[0, 1]^m$, we have

$$\varphi \left(\int_0^1 \cdots \int_0^1 u(t_1, \ldots, t_m) dt_1 \cdots dt_m \right)$$
$$\leq \int_0^1 \cdots \int_0^1 \varphi(u(t_1, \ldots, t_m)) dt_1 \cdots dt_m. \quad (4.5)$$

Indeed, if we choose the exponential function exp as φ and define u by

$$u = 2 \log |F(e^{2\pi\sqrt{-1}t_1}, \ldots, e^{2\pi\sqrt{-1}t_m})|,$$

then, by (4.5), we get $M(F) \leq L_2(F)$.

Further, the inequality (4.5) is proven as follows:
We set

$$c = \int_0^1 \cdots \int_0^1 u(t_1, \ldots, t_m) dt_1 \cdots dt_m.$$

Then the Taylor expansion of φ up to the second order tells us that

$$(x - c)\varphi'(c) \leq \varphi(x) - \varphi(c),$$

so

$$\int_0^1 \cdots \int_0^1 (u - c)\varphi'(c)dt_1 \cdots dt_m \le \int_0^1 \cdots \int_0^1 (\varphi(u) - \varphi(c))dt_1 \cdots dt_m.$$

Note that

$$\int_0^1 \cdots \int_0^1 (\varphi(u) - \varphi(c))dt_1 \cdots dt_m = \int_0^1 \cdots \int_0^1 \varphi(u(t_1, \ldots, t_m))dt_1 \cdots dt_m - \varphi(c)$$

and

$$\int_0^1 \cdots \int_0^1 (u - c)\varphi'(c)dt_1 \cdots dt_m = 0.$$

Thus, the inequality (4.5) is obtained.

3. First, we consider the case $m = 1$. We set $F = a_0 + a_1 X_1 + \cdots + a_d X_1^d$. To prove 3, we may assume $a_d = 1$. Let $F = (X_1 - c_1) \cdots (X_1 - c_d)$ be the decomposition of F. By Jensen's formula (e.g., see [1, Chapter 5, Section 3.1]),

$$\int_0^1 \log |e^{2\pi \sqrt{-1}t} - c| dt = \begin{cases} \log |c| & (\text{if } |c| \ge 1), \\ 0 & (\text{if } |c| < 1), \end{cases}$$

so

$$M(F) = \prod_{i=1}^d \max\{1, |c_i|\}.$$

On the other hand, since $(-1)^j a_j$ is the j-th symmetric form of $c_1, \ldots c_d$, we have $|a_j| \le \binom{d}{j} M(F)$. Since $\binom{d}{j} \le 2^d$, the assertion for $m = 1$ follows.

In general, we prove 3 by induction on m. We assume that $m \ge 2$. We set $d_i = \deg_i(F)$ and

$$F = F_0(X_1, \ldots, X_{m-1}) + F_1(X_1, \ldots, X_{m-1})X_m + \cdots$$
$$+ F_{d_m}(X_1, \ldots, X_{m-1})X^{d_m}.$$

By the case $m = 1$, for fixed $b_1, \ldots, b_{m-1} \in \mathbb{C}$, we have

$$\log(|F_i(b_1, \ldots, b_{m-1})|) \le$$
$$d_m \log(2) + \int_0^1 \log(|F(b_1, \ldots, b_{m-1}, e^{2\pi \sqrt{-1}t_m})|)dt_m.$$

On the other hand, it follows from the induction hypothesis that, for any $i = 0, \ldots, d_m$,

$$\log(|F_i|_\infty) \le (d_1 + \cdots + d_{m-1}) \log(2)$$
$$+ \int_0^1 \cdots \int_0^1 \log(|F_i(e^{2\pi\sqrt{-1}t_1}, \ldots, e^{2\pi\sqrt{-1}t_{m-1}})|) dt_1 \cdots dt_{m-1}.$$

Since b_i is arbitrary, if we set $b_i = e^{2\pi\sqrt{-1}t_i}$, then we have

$$\log(|F|_\infty) = \max_i \{\log(|F_i|_\infty)\} \le (d_1 + \cdots + d_{m-1} + d_m) \log(2) + \log M(F).$$

Thus, the assertion follows.

4. We prove 4 by induction on m. If $m = 1$, then the assertion follows from the following fact:

For any fixed $a, b, c \in \mathbb{Z}_{\ge 0}$,

$$\#\{(i, j) \in \mathbb{Z}_{\ge 0}^2 \mid i + j = c, \ 0 \le i \le a, \ 0 \le j \le b\} \le 1 + \min\{a, b\}. \quad (4.6)$$

Suppose that $m \ge 2$. For $l = 1, \ldots, m$, we set $a_l = \deg_l(F)$ and $b_l = \deg_l(G)$. Further, we set

$$\begin{cases} F = F_0(X, \ldots, X_{m-1}) + \cdots + F_{a_m}(X_1, \ldots, X_{m-1}) X_m^{a_m}, \\ G = G_0(X, \ldots, X_{m-1}) + \cdots + G_{b_m}(X_1, \ldots, X_{m-1}) X_m^{b_m}. \end{cases}$$

Then, for any $l = 1, \ldots, m - 1$, we have $\deg_l(F_i) \le a_l$, $\deg_l(G_j) \le b_l$. Thus, by the induction hypothesis, if we set $c_l = \min\{a_l, b_l\}$ ($l = 1, \ldots, m$), then, by using (4.6), we have

$$|FG|_\infty \le (1 + c_m) \max_{i,j} \{|F_i G_j|_\infty\}$$
$$\le (1 + c_m) \max_{i,j} \{(1 + c_1) \cdots (1 + c_{m-1})|F_i|_\infty |G_j|_\infty\}$$
$$\le (1 + c_m)(1 + c_1) \cdots (1 + c_{m-1})|F|_\infty |G|_\infty.$$

Thus, the assertion follows.

5. If F or G is a constant, then 5 is obvious, so we may assume that $\deg(F) \geq 1$ and $\deg(G) \geq 1$. Then

$$|F|_\infty |G|_\infty$$
$$\leq 2^{m\deg(F)}M(F)2^{m\deg(G)}M(G) \qquad\qquad\qquad \text{(by 3)}$$
$$= 2^{m(\deg(F)+\deg(G))}M(FG)$$
$$\leq 2^{m(\deg(F)+\deg(G))}L_2(FG) \qquad\qquad\qquad \text{(by 2)}$$
$$\leq \left(2^{(\deg(F)+\deg(G))}(\deg(F)+\deg(G)+1)^{1/2}\right)^m |FG|_\infty \quad \text{(by 1)}$$
$$\leq \exp(m(\deg(F)+\deg(G)))|FG|_\infty.$$

Here, we use the formula $M(FG) = M(F)M(G)$ for the third equation. Further, for the last inequality, we use the fact that $2^d(d+1)^{1/2} \leq e^d$ for $d \geq 2$. $\qquad\square$

Now we give a slightly different version of Proposition 4.5,4. This version is easier to work with, and is frequently used in later sections.

Corollary 4.6 *Let $n \geq 2$ and $F_1, \ldots, F_n \in \mathbb{C}[X_1, \ldots, X_m] \setminus \{0\}$. Then*

$$|F_1 \cdots F_n|_\infty \leq |F_1|_\infty \cdots |F_n|_\infty \prod_{i=1}^{n-1}(1+\deg_1(F_i))\cdots(1+\deg_m(F_i)).$$

In particular, $|F_1 \cdots F_n|_\infty \leq |F_1|_\infty \cdots |F_n|_\infty \prod_{i=1}^{n-1}(1+\deg(F_i))^m$.

Proof We prove the assertion by induction on n. In the case $n = 2$, since

$$1 + \min\{\deg_i(F_1), \deg_i(F_2)\} \leq 1 + \deg_i(F_1),$$

the assertion follows from Proposition 4.5,4. We assume that $n \geq 3$. In this case, as before, Proposition 4.5,4 tells us that

$$|F_1 \cdots F_n|_\infty \leq |F_1|_\infty |F_2 \cdots F_n|_\infty (1+\deg_1(F_1))\cdots(1+\deg_m(F_1)).$$

On the other hand, by the induction hypothesis,

$$|F_2 \cdots F_n|_\infty \leq |F_2|_\infty \cdots |F_n|_\infty \prod_{i=2}^{n-1}(1+\deg_1(F_i))\cdots(1+\deg_m(F_i)).$$

Thus, we obtain the assertion. $\qquad\square$

4.2.3 Heights of Polynomials

Let K be a number field. For $F = \sum_{I \in \mathbb{Z}_{\geq 0}^m} a_I X^I \in K[X_1, \ldots, X_m] \setminus \{0\}$, the height $h(F)$ of F is defined by the height of the vector $(a_I)_I$, i.e.,

$$h(F) = h\left((a_I)_{I \in \mathbb{Z}_{\geq 0}^m}\right).$$

Similarly, we define $h^+(F)$ to be $h^+(F) = h^+\left((a_I)_{I \in \mathbb{Z}_{\geq 0}^m}\right)$. Let $F, G \in K[X_1, \ldots, X_m] \setminus \{0\}$. We are going to compare $h(FG)$ with $h(F) + h(G)$. We first consider a special case.

Proposition 4.7 *Let $F, G \in K[X_1, \ldots, X_m] \setminus \{0\}$. If there is $1 \leq i \leq m$ such that $F \in K[X_1, \ldots, X_i]$ and $G \in K[X_{i+1}, \ldots, X_m]$ (i.e., there are no common variables of F and G), then $h(FG) = h(F) + h(G)$.*

Proof The height of F is given by the height of the vector $(a_I)_{I \in \mathbb{Z}_{\geq 0}^i}$ arising from the coefficients of F, and the height of G is given by the height of the vector $(b_J)_{J \in \mathbb{Z}_{\geq 0}^{m-i}}$ arising from the coefficients of G. Since the height of FG is given by the vector $(a_I b_J)$, the assertion follows from Proposition 3.6, 1. $\square$

In general, $F, G \in K[X_1, \ldots, X_m]$ has a common variable. In this case, we have the following estimates. The inequality in Proposition 4.8, 2 is called *Gelfond's inequality*.

Proposition 4.8 *Let $F, G \in K[X_1, \ldots, X_m] \setminus \{0\}$. Let d_i be the degree of F with respect to X_i and let d_i' be the degree of G with respect to X_i. Then we have the following:*

1. $h(FG) \leq h(F) + h(G) + \sum_{i=1}^m \log\left(1 + \min\{d_i, d_i'\}\right)$.
2. $h(F) + h(G) \leq h(FG) + m(\deg(F) + \deg(G))$.

Proof It follows from the definitions of the length $|F|_v$ and the height $h(F)$ that

$$h(F) = \frac{1}{[K : \mathbb{Q}]} \sum_{v \in M_K} \log |F|_v.$$

If v is nonarchimedean, then Proposition 4.4 tells us that $|FG|_v = |F|_v |G|_v$. If v is archimedean, then Proposition 4.5, 4 and 5 tell us that

$$|FG|_v \leq (1 + \min\{d_1, d_1'\}) \cdots (1 + \min\{d_m, d_m'\})|F|_v|G|_v$$

and $|F|_v|G|_v \leq |FG|_v \exp(m(\deg(F) + \deg(G)))$. Thus, we obtain the assertions. $\square$

4.3 Regular Local Ring and Index

In this section, we introduce the index $\text{ind}_x(f;d)$. When f is a one-variable polynomial $F[X]$ over a field F, $a \in F$ and $d = 1$, then the index $\text{ind}_{X-a}(f;1)$ is just the multiplicity of zeros of f at a. The index will be used for Roth's lemma in the next section. Further, for the proof of Faltings's theorem in Chapter 5, we will estimate the index of a section of a suitable line bundle at a suitable x with suitable d.

Let A be a regular local ring and let $\mathfrak{m}$ be its maximal ideal. Let $x_1, \ldots, x_n \in \mathfrak{m}$ be a system of local parameters of A and we write $x = (x_1, \ldots, x_n)$. Further, let F be a field. In this section we assume the following:

Assumption 4.9 $F \subseteq A$ and the homomorphism $F \to A/\mathfrak{m}$ obtained by the composition $F \hookrightarrow A \to A/\mathfrak{m}$ is an isomorphism.

First, we suppose that $(A, \mathfrak{m})$ is complete. Then A is the ring of formal power series with variables $x_1, \ldots, x_n$ over F, i.e., $A = F[\![x_1, \ldots, x_n]\!]$. We refer to [18, Theorem 29.7] for a proof, but it is not difficult to show the assertion.[2]

For $I = (i_1, \ldots, i_n) \in \mathbb{Z}_{\geq 0}^n$, we write $x^I = x_1^{i_1} \cdots x_n^{i_n}$. Then any $f \in A$ is uniquely written as $f = \sum_{I \in \mathbb{Z}_{\geq 0}^n} a_I x^I$ ($a_I \in F$). We say that $a_{(0,\ldots,0)}$, denoted by $f(\mathbf{0})$, is the *constant term* of f.

Let $d_1, \ldots, d_n$ be positive integers and we write $d = (d_1, \ldots, d_n)$. For $I = (i_1, \ldots, i_n) \in \mathbb{Z}_{\geq 0}^n$, we define $|I|_d$ to be

$$|I|_d := \frac{i_1}{d_1} + \cdots + \frac{i_n}{d_n}. \tag{4.7}$$

For $f = \sum_{I \in \mathbb{Z}_{\geq 0}^n} a_I x^I \in A \setminus \{0\}$ ($a_I \in F$), the *index* of f with respect to x and d is defined by $\text{ind}_x(f;d) = \min\{|I|_d \mid a_I \neq 0\}$. When $f = 0$, $\text{ind}_x(f;d)$ is defined to be ∞.

Next, we consider the general case where A need not be complete. In this case, we take the completion $\hat{A}$ of A with respect to $\mathfrak{m}$. Then $\hat{A}$ also satisfies Assumption 4.9 and x forms a system of parameters of $\hat{A}$. For $f \in A$, we define the index of f with respect to x and d by that as an element of $\hat{A}$, and we denote it by $\text{ind}_x(f;d)$.

Proposition 4.10 *For $f, g \in A$, we have the following:*

1. $\text{ind}_x(f + g;d) \geq \min\{\text{ind}_x(f;d), \text{ind}_x(g;d)\}$.

---

[2] Let $\phi: F[\![T_1, \ldots, T_n]\!] \to A$ be the homomorphism given by $\phi(T_i) = x_i$ ($i = 1, \ldots, n$). We can show that ϕ is surjective, and then we obtain the assertion by comparing the dimensions.

2. $\mathrm{ind}_{\boldsymbol{x}}(fg;\boldsymbol{d}) = \mathrm{ind}_{\boldsymbol{x}}(f;\boldsymbol{d}) + \mathrm{ind}_{\boldsymbol{x}}(g;\boldsymbol{d})$.
3. Let k be an integer with $k \geq 1$. Then $\mathrm{ind}_{\boldsymbol{x}}(f;\boldsymbol{d}) = k\,\mathrm{ind}_{\boldsymbol{x}}(f;k\boldsymbol{d})$, where $k\boldsymbol{d} = (kd_1, \ldots, kd_m)$.

Proof We may assume that A is complete. Further, if $f = 0$ or $g = 0$, then the assertions are obvious, so we may assume that $f \neq 0$ and $g \neq 0$. We set

$$\begin{cases} a = \mathrm{ind}_{\boldsymbol{x}}(f;\boldsymbol{d}), \ b = \mathrm{ind}_{\boldsymbol{x}}(g;\boldsymbol{d}), \\ f = \sum_{|I|_{\boldsymbol{d}}=a} a_I \boldsymbol{x}^I + \sum_{|I|_{\boldsymbol{d}}>a} a_I \boldsymbol{x}^I, \\ g = \sum_{|I|_{\boldsymbol{d}}=b} b_I \boldsymbol{x}^I + \sum_{|I|_{\boldsymbol{d}}>b} b_I \boldsymbol{x}^I. \end{cases}$$

1. The assertion follows from the definition of the index.
2. Note that $\left(\sum_{|I|_{\boldsymbol{d}}=a} a_I \boldsymbol{x}^I\right)\left(\sum_{|I|_{\boldsymbol{d}}=b} b_I \boldsymbol{x}^I\right) \neq 0$. Further, if $|I|_{\boldsymbol{d}} = a$ and $|I'|_{\boldsymbol{d}} = b$, then $|I + I'|_{\boldsymbol{d}} = a + b$, which gives the assertion.
3. This is obvious because $|I|_{\boldsymbol{d}} = k|I|_{k\boldsymbol{d}}$. □

Proposition 4.11 *We assume that A is complete. Let $f \in A$. Then we have the following:*

1. Let $\boldsymbol{y} = (y_1, \ldots, y_n)$ be another system of local parameters. If y_i $(i = 1, 2, \ldots, n)$ is written as

$$y_i = a_{i1}x_i + \sum_{k=2}^{\infty} a_{ik}x_i^k \in F[\![x_i]\!] \qquad (a_{i1} \in F^{\times}, a_{i2}, \ldots \in F),$$

then $\mathrm{ind}_{\boldsymbol{x}}(f;\boldsymbol{d}) = \mathrm{ind}_{\boldsymbol{y}}(f;\boldsymbol{d})$.
2. If $u \in A^{\times}$, then $\mathrm{ind}_{\boldsymbol{x}}(uf;\boldsymbol{d}) = \mathrm{ind}_{\boldsymbol{x}}(f;\boldsymbol{d})$.

Proof Clearly we may assume that $f \neq 0$.
  1. First, we claim that x_i is written as

$$x_i = b_{i1}y_i + \sum_{k=2}^{\infty} b_{ik}y_i^k \qquad (b_{i1} \in F^{\times}, b_{i2}, \ldots \in F). \qquad (4.8)$$

Indeed, we substitute $x_i = b_{i1}y_i + \sum_{k=2}^{\infty} b_{ik}y_i^k$ into $y_i = a_{i1}x_i + \sum_{k=2}^{\infty} a_{ik}x_i^k$. Then, comparing the coefficients of y_i^k, we can determine $b_{i1}, b_{i2}, \ldots$ in terms of $a_{i1}, a_{i2}, \ldots$. In particular, $b_{i1} = 1/a_{i1} \neq 0$.
  We set $f = \sum_I a_I \boldsymbol{x}^I = \sum_I a_I' \boldsymbol{y}^I$. We choose $J = (j_1, \ldots, j_n) \in \mathbb{Z}^n$ such that $\mathrm{ind}_{\boldsymbol{x}}(f;\boldsymbol{d}) = |J|_{\boldsymbol{d}}$. Note that $a_J \neq 0$. Further, for $I = (i_1, \ldots, i_m)$, if $i_1 \leq j_1, \ldots, i_n \leq j_n$ and $i_k < j_k$ for some $k \in \{1, \ldots, n\}$, then $a_I = 0$. Thus, if we substitute (4.8) into $x_1, \ldots, x_n$ of f, then we see that $a_J' = a_J b_{11}^{j_1} \cdots b_{n1}^{j_n}$, so that $a_J' \neq 0$. We have $\mathrm{ind}_{\boldsymbol{y}}(f;\boldsymbol{d}) \leq |J|_{\boldsymbol{d}} = \mathrm{ind}_{\boldsymbol{x}}(f;\boldsymbol{d})$.

Applying the same argument with y in place of x, we obtain

$$\text{ind}_x(f;d) \leq \text{ind}_y(f;d),$$

and thus, $\text{ind}_x(f;d) = \text{ind}_y(g;d)$.

2. Let $u_{(0,...,0)}$ be the constant term of u. In the same way as 1, we write $f = \sum_I a_I x^I$, and we take $J = (j_1, \ldots, j_n)$ such that $|J|_d = \text{ind}_x(f;d)$. Then since the coefficient of uf at x^J is $u_{(0,...,0)} a_J \neq 0$, we have

$$\text{ind}_x(uf;d) \leq |J|_d = \text{ind}_x(f;d).$$

By replacing f and u with uf and u^{-1}, respectively, we similarly get

$$\text{ind}_x(u^{-1}(uf);d) \leq \text{ind}_x(uf;d),$$

i.e., $\text{ind}_x(f;d) \leq \text{ind}_x(uf;d)$. Thus, $\text{ind}_x(uf;d) = \text{ind}_x(f;d)$. $\square$

From now on, we assume that the characteristic of F is zero and that A is complete. For $I = (i_1, \ldots, i_n) \in \mathbb{Z}_{\geq 0}^n$, the normalized differential operator ∂_I is defined by

$$\partial_I = \frac{1}{i_1! \cdots i_n!} \left(\frac{\partial}{\partial x_1}\right)^{i_1} \cdots \left(\frac{\partial}{\partial x_n}\right)^{i_n}. \tag{4.9}$$

Proposition 4.12   *1*. $\text{ind}_x(f;d) = \min\{|I|_d \mid \partial_I(f)(0) \neq 0\}$.
2. $\text{ind}_x(\partial_I(f);d) \geq \text{ind}_x(f) - |I|_d$.

Proof   We set $f = \sum_J a_J x^J$. Then

$$\partial_I(f) = \sum_{j_1 \geq i_1, \ldots, j_n \geq i_n} \binom{j_1}{i_1} \cdots \binom{j_n}{i_n} a_{(j_1,...,j_n)} x_1^{j_1 - i_1} \cdots x_n^{j_n - i_n}. \tag{4.10}$$

1. By (4.10), $\partial_I(f)(0) = a_I$, which gives 1.
2. This follows from (4.10). $\square$

Fix a sequence $d = (d_1, \ldots, d_m) \in \mathbb{Z}_{>0}^m$ and $a = (a_1, \ldots, a_m) \in F^m$. Let A be the localization of $F[X_1, \ldots, X_m]$ at the maximal ideal generated by $X_1 - a_1, \ldots, X_n - a_n$. Then $x = (X_1 - a_1, \ldots, X_n - a_n)$ is a regular parameter of A. For $P \in F[X_1, \ldots, X_m]$, the index $\text{ind}_x(P;d)$ is often denoted by $\text{ind}_a(P;d)$, which is called the *index* of P at a with respect to d. By Proposition 4.12, 1, we see that

$$\text{ind}_a(P;d) = \min\{|I|_d \mid \partial_I(P)(a) \neq 0\}.$$

The following proposition is a direct consequence of Propositions 4.10 and 4.12.

Proposition 4.13   *1.* $\operatorname{ind}_a(\partial_I(P); d) \geq \operatorname{ind}_a(P; d) - |I|_d$.
2. $\operatorname{ind}_a(P + Q; d) \geq \min\{\operatorname{ind}_a(P; d), \operatorname{ind}_a(Q; d)\}$.
3. $\operatorname{ind}_a(PQ; d) = \operatorname{ind}_a(P; d) + \operatorname{ind}_a(Q; d)$.
4. *For integer k with $k \geq 1$, $\operatorname{ind}_a(P; d) = k \operatorname{ind}_a(P; kd)$.*

4.4 Roth's Lemma

In his fundamental work on Diophantine approximations of algebraic numbers, Roth proposed the following technical result (cf. [27, Lemma 3C], [28, Theorem 10A]), which is now called Roth's lemma.

Theorem 4.14 (Roth's lemma)   *Fix a sequence $d = (d_1, \ldots, d_m)$ of positive integers, $a = (a_1, \ldots, a_m) \in \overline{\mathbb{Q}}^m$, and a positive number ϵ. Let $P \in \overline{\mathbb{Q}}[X_1, \ldots, X_m]$ be a nonzero polynomial with the following properties:*

1. *The degree of P with respect to the variable X_i is at most d_i for each $i = 1, \ldots, m$.*
2. *$d_{i+1} \leq \epsilon^{2^{m-1}} d_i$ $(i = 1, \ldots, m - 1)$.*
3. *$h(P) + 2md_1 \leq \epsilon^{2^{m-1}} \min_{1 \leq i \leq m}\{d_i h^+(a_i)\}$.*

Then $\operatorname{ind}_a(P; d) \leq 2m\epsilon$. (Note that property 2 is not necessary for the case $m = 1$.)

In this book, since we only need the cases $m = 1$ and $m = 2$ of Roth's lemma, we give the proof for these cases only (see Proposition 4.19 and Theorem 4.20). While Roth's lemma looks technical and complicated, it is surprisingly useful in the theory of Diophantine approximations.

4.4.1 Wronskian

In this subsection, we will see properties of the Wronskian of rational functions for the proof of Roth's lemma. Throughout this subsection, we fix a field K of characteristic zero, and let $K(X)$ be the rational functions field of one variable over K. Note that, for $f \in K(X)$,

$$f \in K \qquad \Longleftrightarrow \qquad f' = 0.$$

We denote the r-th derivative of f by $f^{(r)}$.

Lemma 4.15 *Let $\phi_0, \ldots, \phi_r \in K(X)$. If $\phi_0, \ldots, \phi_r$ are linear independent over K, then*

$$\begin{vmatrix} \phi_0 & \cdots & \phi_r \\ \vdots & \ddots & \vdots \\ \phi_0^{(r)} & \cdots & \phi_r^{(r)} \end{vmatrix} \neq 0.$$

Proof We prove the assertion by induction on r. If $r = 0$, then the assertion is obvious, so we assume $r \geq 1$. We assume the contrary, i.e., the Wronskian is zero. Then

$$\begin{vmatrix} \phi_0 & \cdots & \phi_r \\ \vdots & \ddots & \vdots \\ \phi_0^{(r-1)} & \cdots & \phi_r^{(r-1)} \\ \phi_0^{(t)} & \cdots & \phi_r^{(t)} \end{vmatrix} = 0$$

for all $t = 0, \ldots, r$, so, considering the expansion along the last row, there are $a_0, \ldots, a_r \in K(X)$ such that

$$a_0 \phi_0^{(t)} + \cdots + a_r \phi_r^{(t)} = 0$$

for all $t = 0, \ldots, r$. Note that $a_r \neq 0$ by the induction hypothesis, so we may assume that $a_r = 1$. Since

$$0 = (a_0 \phi_0^{(t)} + \cdots + a_r \phi_r^{(t)})' = a_0' \phi_0^{(t)} + \cdots + a_r' \phi_r^{(t)} + a_0 \phi_0^{(t+1)} + \cdots + a_r \phi_r^{(t+1)},$$

we have $a_0' \phi_0^{(t)} + \cdots + a_{r-1}' \phi_{r-1}^{(t)} = 0$ for all $t = 0, \ldots, r-1$. Thus,

$$\begin{pmatrix} \phi_0 & \cdots & \phi_{r-1} \\ \vdots & \ddots & \vdots \\ \phi_0^{(r-1)} & \cdots & \phi_{r-1}^{(r-1)} \end{pmatrix} \begin{pmatrix} a_0' \\ \vdots \\ a_{r-1}' \end{pmatrix} = \begin{pmatrix} 0 \\ \vdots \\ 0 \end{pmatrix}.$$

It follows from the induction hypothesis that $a_0' = \cdots = a_{r-1}' = 0$. Then $a_0, \ldots, a_{r-1} \in K$. This is a contradiction, because $a_0 \phi_0 + \cdots + a_{r-1} \phi_{r-1} + \phi_r = 0$. $\square$

Let $G \in K[X_1, X_2]$ be a nonzero two-variable polynomial over K. In general, G is not the product of two polynomials with different variables. However, the following proposition ensures that a certain Wronskian of G is indeed the product of two polynomials with different variables. Proposition 4.16 will play a key role in the proof of Roth's lemma.

For $a, b \in \mathbb{Z}_{\geq 0}$, we denote the partial derivative $\partial^{a+b} f / \partial X_1^a \partial X_2^b$ by $\partial^{a,b}(f)$.

Proposition 4.16   *Let $G \in K[X_1, X_2]$ be a nonzero polynomial such that the degree of G with respect to X_i is at most r_i for $i = 1, 2$. Then there is an integer l with the following properties:*

1. *$0 \leq l \leq \min\{r_1, r_2\}$.*
2. *The Wronskian $F = \det\left((\partial^{i,j}(G))_{\substack{0 \leq i \leq l \\ 0 \leq j \leq l}} \right) \neq 0$.*
3. *There is a decomposition $F = UV$ such that $U \in K[X_1]$, $V \in K[X_2]$, the degree of U is at most $(l + 1)r_1$ and the degree of V is at most $(l + 1)r_2$.*

Proof   We write

$$G = \phi_0(X_1)\psi_0(X_2) + \cdots + \phi_l(X_1)\psi_l(X_2)$$

with $\phi_0, \ldots, \phi_l \in K[X_1]$ and $\psi_0, \ldots, \psi_l \in K[X_2]$, where the degrees of $\phi_0, \ldots, \phi_l$ are at most r_1 and the degrees of $\psi_0, \ldots, \psi_l$ are at most r_2. Among all such expressions of G, we choose an expression such that l is the smallest. Note that $0 \leq l \leq \min\{r_1, r_2\}$ and that $\{\phi_0, \ldots, \phi_l\}$ and $\{\psi_0, \ldots, \psi_l\}$ are linearly independent over K. We set

$$U = \det\left((\phi_j^{(i)})_{\substack{0 \leq i \leq l \\ 0 \leq j \leq l}} \right) \quad \text{and} \quad V = \det\left((\psi_j^{(i)})_{\substack{0 \leq i \leq l \\ 0 \leq j \leq l}} \right).$$

Then Lemma 4.15 tells us that $U \neq 0$ and $V \neq 0$. Further, as $(\partial^{i,j})(G) = \sum_{s=0}^{l} \phi_s^{(i)} \psi_s^{(j)}$, we have $F = UV$. $\square$

In the following, we assume that K is a number field. As an application of Propositions 4.4 and 4.5, we give an estimate of the length and the height of the Wronskian in Proposition 4.16.

Corollary 4.17   *Under the setting of Proposition 4.16, we have*

$$|F|_v \leq \begin{cases} |G|_v^{l+1} & (\text{if } v \in M_K^{\text{fin}}), \\[2mm] (l+1)!\,((r_1+1)(r_2+1))^{l+1}\, 2^{(l+1)(r_1+r_2)} |G|_v^{l+1} & (\text{if } v \in M_K^{\infty}). \end{cases}$$

Proof   Note that

$$\partial^{a,b}(X_1^i X_2^j) = \begin{cases} \binom{i}{a}\binom{j}{b} X_1^{i-a} X_2^{j-b} & (\text{if } i \geq a \text{ and } j \geq b), \\ 0 & (\text{otherwise}). \end{cases}$$

It follows from $\binom{a}{b} \leq 2^a$ that

$$|\partial^{a,b}(G)|_v \leq \begin{cases} |G|_v & (\text{if } v \in M_K^{\text{fin}}), \\ 2^{r_1+r_2}|G|_v & (\text{if } v \in M_K^{\infty}). \end{cases}$$

Then, by Proposition 4.5, 4, each term in the expansion of the determinant $\det\left(\left(\partial^{i,j}(G) \right)_{\substack{0 \le i \le l \\ 0 \le j \le l}} \right)$ is bounded above by

$$\begin{cases} |G|_v^{l+1} & (\text{if } v \in M_K^{\text{fin}}), \\ ((r_1+1)(r_2+1))^{l+1} \, 2^{(l+1)(r_1+r_2)} |G|_v^{l+1} & (\text{if } v \in M_K^{\infty}). \end{cases}$$

We obtain the assertion. $\qquad\square$

Corollary 4.18 *Under the setting of Proposition 4.16, we have*

$$h(F) \le (l+1)\left(h(G) + 2(r_1+r_2)\log(2) + l\right).$$

Proof By Corollary 4.17,

$$\begin{aligned} h(F) \le (l+1)h(G) + (l+1)(r_1+r_2)\log(2) \\ + (l+1)\log((r_1+1)(r_2+1)) + \log((l+1)!). \end{aligned}$$

Since $1 + a \le 2^a$ for any $a \in \mathbb{Z}_{\ge 0}$ and $\log((l+1)!) \le (l+1)\log(l+1) \le (l+1)l$ for any $l \in \mathbb{Z}_{\ge 0}$, we obtain the assertion. $\qquad\square$

4.4.2 Roth's Lemma for $m = 1$ and $m = 2$

In this subsection, we give the proof of Roth's lemma for $m = 1$ and $m = 2$.

Proposition 4.19 (Roth's lemma for $m = 1$) *Fix $d \in \mathbb{Z}_{>0}$, $a \in \overline{\mathbb{Q}}$ and a positive number ϵ. Let $P \in \overline{\mathbb{Q}}[X]$ be a nonzero polynomial such that $\deg(P) \le d$ and $h(P) + 2d \le \epsilon d h^+(a)$. Then $\mathrm{ind}_a(P; d) \le \epsilon$.*

Proof We write $P(X) = (X-a)^i Q(X)$ such that $i \in \mathbb{Z}_{\ge 0}$ and $Q(X) \in \overline{\mathbb{Q}}[X]$ with $Q(a) \neq 0$. By Proposition 4.8, we have $h((X-a)^i) + h(Q) \le h(P) + \deg(P)$. Since

$$ih^+(a) = h(X^i + (-a)^i) \le h\left(\sum_{s=0}^{i} \binom{i}{s}(-a)^{i-s} X^s \right) = h\left((X-a)^i \right),$$

our assumptions tell us that

$$ih^+(a) \le h(P) + \deg(P) \le h(P) + d \le \epsilon d h^+(a) - d \le \epsilon d h^+(a).$$

Since the assumption $h(P) + 2d \le \epsilon d h^+(a)$ implies that $h^+(a) > 0$, we have $\mathrm{ind}_a(P; d) = \frac{i}{d} \le \epsilon$. $\qquad\square$

Theorem 4.20 (Roth's lemma for $m = 2$) *Fix $d_1, d_2 \in \mathbb{Z}_{>0}$, $a_1, a_2 \in \overline{\mathbb{Q}}$ and a positive number ϵ. Let $P \in \overline{\mathbb{Q}}[X_1, X_2]$ be a nonzero polynomial with the following properties:*

1. *The degree of P with respect to X_i is at most d_i for each $i = 1, 2$.*
2. *$d_2 \leq \epsilon^2 d_1$.*
3. *$h(P) + 4d_1 \leq \epsilon^2 \min_{i=1,2}\{d_i h^+(a_i)\}$.*

Then $\mathrm{ind}_a(P; d) \leq 4\epsilon$, where $d = (d_1, d_2)$ and $a = (a_1, a_2)$.

First, we show the following elementary inequality:

Lemma 4.21 *Let $\alpha \in [0, \infty)$, let d be a positive integer, and let l be an integer with $0 \leq l \leq d$. Then*

$$\sum_{j=0}^{l}\left(\max\left\{\alpha - \frac{j}{d}, 0\right\}\right) \geq (l+1)\min\left\{\frac{\alpha}{2}, \frac{\alpha^2}{4}\right\}.$$

Proof First, suppose that $\alpha \geq l/d$. Then

$$\sum_{j=0}^{l}\left(\max\left\{\alpha - \frac{j}{d}, 0\right\}\right) = (l+1)\alpha - \frac{l(l+1)}{2d}$$

$$\geq (l+1)\alpha - \frac{(l+1)\alpha}{2} = (l+1)\frac{\alpha}{2}.$$

Next, suppose that $0 \leq \alpha < l/d$. We take an integer s with $s/d \leq \alpha < (s+1)/d$ and $0 \leq s \leq l-1$. Since $\max\left\{\alpha - \frac{j}{d}, 0\right\} = 0$ for $j \geq s+1$, the above argument gives

$$\sum_{j=0}^{l}\left(\max\left\{\alpha - \frac{j}{d}, 0\right\}\right) \geq (s+1)\frac{\alpha}{2}.$$

Note that $s + 1 > d\alpha$ and $d/(d+1) \geq 1/2$ (because $d \geq 1$). Then

$$\sum_{j=0}^{l}\left(\max\left\{\alpha - \frac{j}{d}, 0\right\}\right) \geq \frac{d}{2}\alpha^2 = (d+1)\frac{d}{2(d+1)}\alpha^2$$

$$\geq (l+1)\frac{d}{2(d+1)}\alpha^2 \geq (l+1)\frac{\alpha^2}{4},$$

as desired.   □

Proof of Theorem 4.20 We take a number field K such that

$$P \in K[X_1, X_2] \quad \text{and} \quad a \in K^2.$$

If $P(\boldsymbol{a}) \neq 0$, then $\mathrm{ind}_{\boldsymbol{a}}(P;\boldsymbol{d}) = 0$, and the assertion is obvious. We may assume that $P(\boldsymbol{a}) = 0$. We divide the proof into six steps.

Step 1. Since $\mathrm{ind}_{\boldsymbol{a}}(P;\boldsymbol{d}) \leq 2$, we may assume that $\epsilon \leq 1/2$. By Proposition 4.16, there is an integer l with $0 \leq l \leq \min\{d_1, d_2\}$ such that the Wronskian

$$F = \det\left(\left(\partial^{i,j}(P)\right)_{\substack{0 \leq i \leq l \\ 0 \leq j \leq l}}\right)$$

is nonzero and decomposes into a product $F = UV$, where $U \in K[X_1]$, $V \in K[X_2]$, $\deg(U) \leq (l+1)d_i$ and $\deg(V) \leq (l+1)d_2$.

Step 2. We estimate $\mathrm{ind}_{\boldsymbol{a}}(P;\boldsymbol{d})$. By the assumption 2 of Theorem 4.20, we have $l \leq d_2 \leq \epsilon^2 d_1$. It follows from Proposition 4.13,1 that

$$\mathrm{ind}_{\boldsymbol{a}}(\partial^{i,j}(P);\boldsymbol{d}) \geq \mathrm{ind}_{\boldsymbol{a}}(P;\boldsymbol{d}) - \left(\frac{i}{d_1} + \frac{j}{d_2}\right) \geq \mathrm{ind}_{\boldsymbol{a}}(P;\boldsymbol{d}) - \frac{d_2}{d_1} - \frac{j}{d_2}.$$

Since $\mathrm{ind}_{\boldsymbol{a}}(\partial^{i,j}(P);\boldsymbol{d}) \geq 0$, we obtain

$$\mathrm{ind}_{\boldsymbol{a}}(\partial^{i,j}(P);\boldsymbol{d}) \geq \max\left\{\mathrm{ind}_{\boldsymbol{a}}(P;\boldsymbol{d}) - \frac{j}{d_2}, 0\right\} - \frac{d_2}{d_1}.$$

Then Proposition 4.13,2–3 together with Lemma 4.21 tell us that

$$\mathrm{ind}_{\boldsymbol{a}}(F;\boldsymbol{d}) \geq \sum_{j=0}^{l}\left(\max\left\{\mathrm{ind}_{\boldsymbol{a}}(P;\boldsymbol{d}) - \frac{j}{d_2}, 0\right\}\right) - (l+1)\frac{d_2}{d_1}$$

$$\geq (l+1)\min\left\{\frac{\mathrm{ind}_{\boldsymbol{a}}(P;\boldsymbol{d})}{2}, \frac{\mathrm{ind}_{\boldsymbol{a}}(P;\boldsymbol{d})^2}{4}\right\} - (l+1)\frac{d_2}{d_1}.$$

Thus, since $\mathrm{ind}_{\boldsymbol{a}}(F;\boldsymbol{d}) = \mathrm{ind}_{\boldsymbol{a}}(U;\boldsymbol{d}) + \mathrm{ind}_{\boldsymbol{a}}(V;\boldsymbol{d})$ and $d_2/d_1 \leq \epsilon^2$, we have

$$\min\left\{\frac{\mathrm{ind}_{\boldsymbol{a}}(P;\boldsymbol{d})}{2}, \frac{\mathrm{ind}_{\boldsymbol{a}}(P;\boldsymbol{d})^2}{4}\right\} \leq \frac{1}{l+1}\mathrm{ind}_{\boldsymbol{a}}(U;\boldsymbol{d}) + \frac{1}{l+1}\mathrm{ind}_{\boldsymbol{a}}(V;\boldsymbol{d}) + \epsilon^2.$$

$$(4.11)$$

Using the case $m = 1$, we are going to give upper bounds of $\mathrm{ind}_{\boldsymbol{a}}(U;\boldsymbol{d})$ and $\mathrm{ind}_{\boldsymbol{a}}(V;\boldsymbol{d})$, which will then give an upper bound of $\mathrm{ind}_{\boldsymbol{a}}(P;\boldsymbol{d})$.

Step 3. In this step, we claim

$$h(U) + h(V) \leq (l+1)(h(P) + 2d_1). \tag{4.12}$$

Indeed, since $F = UV$, and U and V have no common variables, Proposition 4.7 tells us that $h(F) = h(U) + h(V)$. Further, by Corollary 4.18, we have

$$h(F) \leq (l+1)(h(P) + 2(d_1 + d_2)\log(2) + l).$$

Since $l \le d_2 \le \epsilon^2 d_1$, we get

$$h(F) \le (l+1)\left(h(P) + \left(2(1+\epsilon^2)\log(2) + \epsilon^2\right)d_1\right).$$

Thus, to show (4.12), it suffices to verify

$$2(1+\epsilon^2)\log(2) + \epsilon^2 \le 2.$$

Using $\log(2) = 0.69\ldots < 7/10$ and $\epsilon^2 \le 1/4$, we have

$$2(1+\epsilon^2)\log(2) + \epsilon^2 \le 2(1+1/4)\log(2) + 1/4 < 2(1+1/4)(7/10) + 1/4 = 2.$$

This completes the proof of the claim (4.12).

Step 4. Since $h(U) \ge 0$ and $h(V) \ge 0$, the estimate (4.12) tells us that

$$h(U) \le (l+1)(h(P) + 2d_1) \quad \text{and} \quad h(V) \le (l+1)(h(P) + 2d_1).$$

Then, it follows from the assumption 2 of Theorem 4.20 that

$$h(U) + 2(l+1)d_1 \le (l+1)(h(P) + 4d_1)$$
$$\le \epsilon^2 \min_{i=1,2}\{(l+1)d_i h^+(a_i)\} \le \epsilon^2(l+1)d_1 h^+(a_1)$$

and

$$h(V) + 2(l+1)d_2 \le h(V) + 2(l+1)d_1 \le (l+1)(h(P) + 4d_1)$$
$$\le \epsilon^2 \min_{i=1,2}\{(l+1)d_i h^+(a_i)\} \le \epsilon^2(l+1)d_2 h^+(a_2).$$

Applying Proposition 4.19 to U and V with $\epsilon(U) = \epsilon^2$ for U and $\epsilon(V) = \epsilon^2$ for V, we have

$$\mathrm{ind}_{a_1}(U;(l+1)d_1) \le \epsilon^2 \quad \text{and} \quad \mathrm{ind}_{a_2}(V;(l+1)d_2) \le \epsilon^2.$$

Using Proposition 4.13,4, we have

$$\mathrm{ind}_a(U;\boldsymbol{d}) \le (l+1)\epsilon^2 \quad \text{and} \quad \mathrm{ind}_a(V;\boldsymbol{d}) \le (l+1)\epsilon^2.$$

Then the inequality (4.11) gives

$$\min\left\{\frac{\mathrm{ind}_a(P;\boldsymbol{d})}{2}, \frac{\mathrm{ind}_a(P;\boldsymbol{d})^2}{4}\right\} \le \epsilon^2 + \epsilon^2 + \epsilon^2 \le 3\epsilon^2.$$

Step 5. In the final step, we consider two cases. First, suppose that $\mathrm{ind}_a(P;\boldsymbol{d})/2 \le \mathrm{ind}_a(P;\boldsymbol{d})^2/4$. Then $\mathrm{ind}_a(P;\boldsymbol{d})/2 \le 3\epsilon^2$. Since $\epsilon \le 1/2$, we have

$$\mathrm{ind}_a(P;\boldsymbol{d}) \le 6\epsilon^2 = (2\epsilon)3\epsilon \le 3\epsilon < 4\epsilon.$$

Next, suppose that $\text{ind}_a(P;\mathbf{d})/2 > \text{ind}_a(P;\mathbf{d})^2/4$. Then $\text{ind}_a(P;\mathbf{d}) \leq \sqrt{12}\epsilon$. Since $\sqrt{12} < 4$, we have $\text{ind}_a(P;\mathbf{d}) \leq 4\epsilon$, which completes the proof of the theorem. $\qquad\square$

4.5 Norms of Invertible Sheaves

In this section, we give the definition and properties of the norms of invertible sheaves. Details can be found in, for example, [20, Section 1.5]. Let $\pi : X \to Y$ be a finite and surjective morphism of algebraic schemes over a field k, and let L be an invertible sheaf on X. By [20, Lemma 1.15], there is an affine open covering $Y = \bigcup_{i=1}^{N} Y_i$ such that, for each i, we can find a local basis ω_i of L over $\pi^{-1}(Y_i)$. Here, we consider the following two assumptions:

Assumption 4.22 π is flat.

Assumption 4.23 X and Y are normal algebraic varieties and the function field E of X is separable over the function field F of Y.

We are going to define the norms of invertible sheaves under one of the above two assumptions.

4.5.1 Case under Assumption 4.22

Here, we assume Assumption 4.22, i.e., π is flat. We begin with the following lemma:

Lemma 4.24 *Let A be a local ring, and let B be an A-algebra such that B is finitely generated over A as an A-module. We assume that B is flat over A, i.e., B is a free A-module of finite rank.*

1. Let $\{x_1, \ldots, x_n\}$ be a basis of B as a free A-module. For $b \in B$, we set

$$bx_j = a_{1j}x_1 + \cdots + a_{nj}x_n.$$

Then $\det(a_{ij})$, denoted by $\det(b\cdot)$, does not depend on the choice of $\{x_1, \ldots, x_n\}$.

2. For any $b, b' \in B$, we have $\det(bb'\cdot) = \det(b\cdot)\det(b'\cdot)$.

Proof 1. Let $\{x'_1,\ldots,x'_n\}$ be another free basis of B over A. We set $x'_j = \sum_{i=1}^n c_{ij}x_i$, $x_j = \sum_{i=1}^n c'_{ij}x'_i$ and $bx'_j = \sum_{i=1}^n a'_{ij}x'_i$. Then

$$\sum_{m=1}^n a'_{mj}x'_m = bx'_j = b\left(\sum_{i=1}^n c_{ij}x_i\right) = \sum_{i=1}^n c_{ij}(bx_i) = \sum_{i=1}^n c_{ij}\sum_{l=1}^n a_{li}x_l$$

$$= \sum_{i=1}^n\sum_{l=1}^n c_{ij}a_{li}\sum_{m=1}^n c'_{ml}x'_m = \sum_{m=1}^n\left(\sum_{l=1}^n\sum_{i=1}^n c'_{ml}a_{li}c_{ij}\right)x'_m.$$

Thus, $a'_{mj} = \sum_{l=1}^n\sum_{i=1}^n c'_{ml}a_{li}c_{ij}$. It follows that

$$(a'_{ij}) = (c'_{ij})(a_{ij})(c_{ij}).$$

Since (c'_{ij}) is the inverse matrix of (c_{ij}), the assertion follows.

2. We set $bx_j = \sum_{i=1}^n a_{ij}x_i$ and $b'x_j = \sum_{i=1}^n e_{ij}x_i$. Then

$$bb'x_j = b\left(\sum_{i=1}^n e_{ij}x_i\right) = \sum_{i=1}^n e_{ij}bx_i = \sum_{i=1}^n e_{ij}\sum_{l=1}^n a_{li}x_l = \sum_{l=1}^n\left(\sum_{i=1}^n a_{li}e_{ij}\right)x_l.$$

Thus, the matrix representation of bb' is $(a_{ij})(e_{ij})$, and 2 follows. $\square$

For $b \in H^0(X,\mathcal{O}_X)$, since X is flat over Y, the determinant of the multiplication homomorphism

$$b\cdot\colon \mathcal{O}_X \to \mathcal{O}_X \quad (x \mapsto bx)$$

by b is defined locally with respect to Y. By Lemma 4.24, 1, it gives rise to an element in $H^0(Y,\mathcal{O}_Y)$, which we denote by $\mathrm{Norm}_\pi(b)$. Further, for $b,b' \in H^0(X,\mathcal{O}_X)$, Lemma 4.24,2 tells us that

$$\mathrm{Norm}_\pi(bb') = \mathrm{Norm}_\pi(b)\,\mathrm{Norm}_\pi(b'). \tag{4.13}$$

It follows that, if $b \in H^0(X,\mathcal{O}_X^\times)$, then $\mathrm{Norm}_\pi(b) \in H^0(Y,\mathcal{O}_Y^\times)$.

Suppose that an invertible sheaf L on X is given. Then there is an affine open covering $Y = \bigcup_{i=1}^N Y_i$ such that, for each i, we can find a local basis ω_i of L over $\pi^{-1}(Y_i)$. For each i,j, we take $g_{ij} \in H^0(\pi^{-1}(Y_i \cap Y_j),\mathcal{O}_X^\times)$ such that $\omega_j = g_{ij}\omega_i$ over $\pi^{-1}(Y_i \cap Y_j)$. Since $g_{il} = g_{ij}g_{jl}$, by (4.13), we have $\mathrm{Norm}_\pi(g_{il}) = \mathrm{Norm}_\pi(g_{ij})\,\mathrm{Norm}_\pi(g_{jl})$. Thus, $\{\mathrm{Norm}_\pi(g_{ij})\}$ gives rise to an invertible sheaf on Y, which we denote by $\mathrm{Norm}_\pi(L)$.

For $s \in H^0(X,L)$, we set $s = f_i\omega_i$ over $\pi^{-1}(Y_i)$, where f_i is an element of $H^0(\pi^{-1}(Y_i),\mathcal{O}_X)$. Since $f_i = g_{ij}f_j$, it follows from (4.13) that

$$\mathrm{Norm}_\pi(f_i) = \mathrm{Norm}_\pi(g_{ij})\,\mathrm{Norm}_\pi(f_j).$$

Thus, $\{\mathrm{Norm}_\pi(f_i)\}$ gives rise to an element of $H^0(Y,\mathrm{Norm}_\pi(L))$, which we denote by $\mathrm{Norm}_\pi(s)$.

Proposition 4.25 *Let L, L' be invertible sheaves on X, and let $s \in H^0(X, L)$ and $s' \in H^0(X, L')$. Then we have*

$$\begin{cases} \mathrm{Norm}_\pi (L \otimes L') = \mathrm{Norm}_\pi (L) \otimes \mathrm{Norm}_\pi (L'), \\ \mathrm{Norm}_\pi (s \otimes s') = \mathrm{Norm}_\pi (s) \otimes \mathrm{Norm}_\pi (s'). \end{cases}$$

Proof Let $L'' := L \otimes L'$ and $Y = \bigcup_{i=1}^N Y_i$ be an affine open covering of Y such that L and L' have local bases ω_i and ω_i' over $\pi^{-1}(Y_i)$, respectively. We define $g_{ij}, g_{ij}' \in H^0(\pi^{-1}(Y_i \cap Y_j), \mathcal{O}_X^\times)$ by the equations $\omega_j = g_{ij}\omega_i$ and $\omega_j' = g_{ij}'\omega_i'$. Then $\mathrm{Norm}_\pi (L)$ and $\mathrm{Norm}_\pi (L')$ are defined by $\{\mathrm{Norm}(g_{ij})\}$ and $\{\mathrm{Norm}(g_{ij}')\}$. On the other hand, L'' and $\mathrm{Norm}_\pi (L'')$ are given by $\{g_{ij}g_{ij}'\}$ and $\{\mathrm{Norm}_\pi (g_{ij}g_{ij}')\}$, respectively. Since

$$\mathrm{Norm}_\pi (g_{ij}g_{ij}') = \mathrm{Norm}_\pi (g_{ij}) \mathrm{Norm}_\pi (g_{ij}'),$$

the first equation follows. If we set $s = s_i\omega_i$ and $s' = s_i'\omega_i'$ on $\pi^{-1}(Y_i)$, then $s \otimes s' = s_i s_i'(\omega_i \otimes \omega_i')$. Since $\mathrm{Norm}(s_i s_i') = \mathrm{Norm}(s_i) \mathrm{Norm}(s_i')$, the second equation follows. $\qquad\square$

4.5.2 Case under Assumption 4.23

In this subsection, we assume Assumption 4.23, i.e., X and Y are normal algebraic varieties and the function field E of X is separable over the function field F of Y. Let D be a Cartier divisor on X.

Lemma 4.26 *There is an affine open covering $Y = \bigcup_{i=1}^N Y_i$ such that, for each i, we can find a local equation f_i of D over $\pi^{-1}(Y_i)$.*

Proof As we explained in the beginning of this section, there is an affine open covering $Y = \bigcup_{i=1}^N Y_i$ such that, for each i, we can find a local basis of ϕ_i of $\mathcal{O}_X(D)$. Note that $\phi_i \in H^0(\pi^{-1}(Y_i), \mathcal{O}_X(D))$, so $\phi_i \in E^\times$. It suffices to show that ϕ_i^{-1} gives a local equation of D. This is a local problem. Let V be a Zariski open set such that $V \subseteq \pi^{-1}(Y_i)$ and a local equation of D over V is given by f. Then there is $u \in \mathcal{O}_V^\times$ with $f^{-1} = u\phi_i$ on V, i.e., $\phi_i^{-1} = uf$. $\qquad\square$

We use the same notation in the beginning of this section. In particular, L is an invertible sheaf on X, $\{Y_i\}_{i=1}^N$ is an affine open covering of Y, and ω_i is a local basis of L over $\pi^{-1}(Y_i)$. In the same way as in the case under Assumption 4.22, if we define g_{ij} on $\pi^{-1}(Y_i \cap Y_j)$ by $\omega_j = g_{ij}\omega_i$, then $g_{ij} \in \mathcal{O}_{\pi^{-1}(Y_i \cap Y_j)}^\times$. Thus, Lemma 2.2 tells us that $\mathrm{Norm}(g_{ij}) \in \mathcal{O}_{Y_i \cap Y_j}^\times$. Further, since $g_{kj} = g_{ki}g_{ij}$, we obtain $\mathrm{Norm}(g_{kj}) = \mathrm{Norm}(g_{ki}) \mathrm{Norm}(g_{ij})$.

It follows that $\{\mathrm{Norm}(g_{ij})\}$ gives rise to an invertible sheaf on Y, which is denoted by $\mathrm{Norm}_\pi(L)$.

Here, we consider the Cartier divisor D. Since $f_j/f_i \in \mathcal{O}^\times_{\pi^{-1}(Y_i \cap Y_j)}$, we have $\mathrm{Norm}(f_i)/\mathrm{Norm}(f_j) \in \mathcal{O}^\times_{Y_i \cap Y_j}$. Thus, $\{\mathrm{Norm}(f_i)\}$ gives a Cartier divisor on Y, which is denoted by $\mathrm{Norm}_\pi(D)$.

For a rational section s of L, we define $s_i \in E$ by $s|_{\pi^{-1}(Y_i)} = s_i \omega_i$. Then, since $s_i = g_{ij} s_j$, we have $\mathrm{Norm}(s_i) = \mathrm{Norm}(g_{ij})\,\mathrm{Norm}(s_j)$ over $Y_i \cap Y_j$, so $\{\mathrm{Norm}(s_i)\}$ defines the rational section of $\mathrm{Norm}_\pi(L)$. It is denoted by $\mathrm{Norm}_\pi(s)$. If $s \in H^0(X, L)$, then by Lemma 2.2 $\mathrm{Norm}(s_i) \in \mathcal{O}_{Y_i}$ because $s_i \in \mathcal{O}_{\pi^{-1}(Y_i)}$. Thus, $\mathrm{Norm}_\pi(s) \in H^0(X, \mathrm{Norm}_\pi(L))$. Further, by our construction, we have

$$\mathrm{Norm}_\pi(\mathrm{div}(s)) = \mathrm{div}(\mathrm{Norm}_\pi(s)). \qquad (4.14)$$

(Here, the left-hand side is the norm of the Cartier divisor $\mathrm{div}(s)$ on X, and the right-hand side is the Cartier divisor of the norm of a rational section s.)

Proposition 4.27 *1. Let L' be another invertible sheaf on X and s' be a rational section of L'. Then $\mathrm{Norm}_\pi(L \otimes L') = \mathrm{Norm}_\pi(L) \otimes \mathrm{Norm}_\pi(L')$ and $\mathrm{Norm}_\pi(s \otimes s') = \mathrm{Norm}_\pi(s) \otimes \mathrm{Norm}_\pi(s')$.*
2. If $s \in H^0(X, L) \setminus \{0\}$, then

$$\pi^*(\mathrm{Norm}_\pi(s)) \otimes s^{-1} \in H^0(X, \pi^*(\mathrm{Norm}_\pi(L)) \otimes L^{-1}).$$

Proof 1. It can be proved in the same way as Proposition 4.25.

2. Let $\{Y_i\}$ be an affine open covering of Y. To prove 2, it suffices to show that $\pi^*(\mathrm{Norm}_\pi(s)) \otimes s^{-1} \in H^0(\pi^{-1}(Y_i), \pi^*(\mathrm{Norm}_\pi(L)) \otimes L^{-1})$ for all i. Thus, we may assume that $X = \mathrm{Spec}(B)$, $Y = \mathrm{Spec}(A)$, and $L = B$. In this case, the assertion follows from Lemma 2.2. $\qquad \square$

Finally, we observe that, if s is a nonzero rational section of L, then

$$\pi_* \mathrm{div}(s) = \mathrm{div}(\mathrm{Norm}_\pi(s)) = \mathrm{Norm}_\pi(\mathrm{div}(s)), \qquad (4.15)$$

where $\pi_* \mathrm{div}(s)$ is the push forward[3] of $\mathrm{div}(s)$.

Indeed, since the second equality is exactly (4.14), we consider the first equality. By Proposition 4.27, we may assume that s is a nonzero global section of L. Since it suffices to show the first equality at each codimension one point $y \in Y$, we may further assume that $Y = \mathrm{Spec}(\mathcal{O}_{Y,y})$ and

---

[3] If D is a prime divisor, then the push forward $\pi_* D$ of D is defined as $\pi_* D = [\kappa(P) : \kappa(\pi(P))]\pi(D)$, where P is the generic point of D, and $\kappa(P)$ and $\kappa(\pi(P))$ are the residue fields of P and $\pi(P)$, respectively. In the general case, we extend the definition by the linearity of π_*.

$X = \pi^{-1}(Y) = \mathrm{Spec}(R)$, where R is a finite $\mathcal{O}_{Y,y}$-algebra. Then the assertion can be reduced to [20, Lemma 1.11 (2)].

4.6 Height of Norm

We have discussed norms in Sections 2.1 and 4.5. In this section, we estimate heights of norms (see Proposition 4.30). The content of this section may look complicated, but it is only repetition of simple calculations. We begin with the following lemma:

Lemma 4.28 *Let K be a number field and let $K[Y, Z]$ be the polynomial ring of two variables Y and Z over K. Let $F \in K[Y, Z]$ be a polynomial of degree N of the form*

$$F = Z^N + a_{N-1}(Y)Z^{N-1} + \cdots + a_0(Y) \quad (a_i(Y) \in K[Y], \ \deg(a_i) \le N - i).$$

Let $k[y, z] = K[Y, Z]/(F)$ be the quotient ring, where y, z denote the classes of Y, Z. Note that $\{1, z, \ldots, z^{N-1}\}$ forms a free basis of $K[y, z]$ as a $K[y]$-module. Also, since $K[y]$ is a polynomial ring of one variable y, for any $l \ge 0$, there are unique $\beta_{l,i}(Y) \in K[Y]$ $(i = 0, \ldots, N - 1)$ such that

$$z^l = \sum_{0 \le i < N} \beta_{l,i}(y)z^i.$$

Then we have the following:

1. *$\beta_{l,i}(Y)$ is a polynomial of degree at most $(l - i)$. Further, if $a_0, \ldots, a_{N-1} \in O_K[Y]$, then $\beta_{l,i}(Y) \in O_K[Y]$.*
2. *For each $v \in M_K$, there is a nonnegative constant c_v depending only on $N, a_0, \ldots, a_{N-1}$ and v such that*

$$|\beta_{l,i}|_v \le \exp(c_v l)$$

for all l, i. Further, we can take $c_v = 0$ except for finitely many v.

Proof 1. First, since

$$z^{l+1} = z \sum_{i=0}^{N-1} \beta_{l,i}z^i = \beta_{l,N-1}z^N + \sum_{i=1}^{N-1} \beta_{l,i-1}z^i$$

$$= -\beta_{l,N-1} \sum_{i=0}^{N-1} a_i z^i + \sum_{i=1}^{N-1} \beta_{l,i-1}z^i$$

$$= \sum_{i=1}^{N-1} (-\beta_{l,N-1}a_i + \beta_{l,i-1})z^i - \beta_{l,N-1}a_0,$$

we have

$$\begin{cases} \beta_{l+1,i} = -\beta_{l,N-1}a_i + \beta_{l,i-1} & (1 \le i \le N-1), \\ \beta_{l+1,0} = -\beta_{l,N-1}a_0. \end{cases} \qquad (4.16)$$

We prove the assertion $\deg(\beta_{l,i}(Y)) \le l - i$ by induction on l. If $0 \le l < N$, then the assertion is obvious, because

$$\beta_{l,i} = \begin{cases} 1 & (\text{if } i = l), \\ 0 & (\text{if } i \ne l). \end{cases}$$

(Note that the degree of the zero polynomial is $-\infty$. See N4 on p. 163). Further, since $\beta_{N,i} = -a_i$, it holds for $l = N$. We assume that $l \ge N$. By the hypothesis of the induction,

$$\begin{cases} \deg(\beta_{l,N-1}) \le l - N + 1, & \deg(a_0) \le N, \\ \deg(\beta_{l,N-1}) \le l - N + 1, & \deg(a_i) \le N - i, \quad \deg(\beta_{l,i-1}) \le l - i + 1 \end{cases}$$

for any $1 \le i \le N - 1$. Thus, by (4.16), we have

$$\deg(\beta_{l+1,0}) \le l + 1 \quad \text{and} \quad \deg(\beta_{l+1,i}) \le l - i + 1,$$

as desired. Further, if $a_i \in O_K[Y]$ for all i, then, by (4.16), we see $\beta_{l,i}(Y) \in O_K[Y]$.

2. For $v \in M_K$, we set

$$H_v = \max\{1, |a_0|_v, \ldots, |a_{N-1}|_v\},$$

where $|a_i|_v$ is the length of the polynomial $a_i = a_i(Y)$ with respect to v (see Section 4.2). Then, except for finitely many v, $H_v = 1$. Further, if we set

$$c_v = \begin{cases} \log(H_v) & (\text{if } v \in M_K^{\text{fin}}), \\ \log((N+1)H_v + 1) & (\text{if } v \in M_K^{\infty}), \end{cases}$$

then $c_v = 0$ except for finitely many v. We prove the inequality on $|\beta_{l,i}|_v$ by induction on l. If $0 \le l \le N$, then $\beta_{l,i} \in \{0, 1, -a_0, \ldots, -a_{N-1}\}$, so the assertion is obvious. We may assume $l \ge N$.

First, we consider the case where v is nonarchimedean. By Gauss's lemma (Proposition 4.4) and (4.16), for any $i > 0$, we have

$$\begin{aligned} |\beta_{l+1,i}|_v &\le \max\{|\beta_{l,N-1}a_i|_v, |\beta_{l,i-1}|_v\} = \max\{|\beta_{l,N-1}|_v|a_i|_v, |\beta_{l,i-1}|_v\} \\ &\le \max\{\exp(c_v l)H_v, \exp(c_v l)\} = \exp(c_v l)H_v \\ &= \exp(c_v l)\exp(c_v) = \exp(c_v(l+1)). \end{aligned}$$

If $i = 0$, then, since $\beta_{l+1,0} = -\beta_{l,0}a_0$, we can show the assertion in the same way as above.

Next, we consider the case where v is archimedean. If $i > 0$, then, since $\beta_{l,N-1}, a_i$ are polynomials of one variable and $\deg(a_i) \leq N$, Corollary 4.6 tells us that

$$
\begin{aligned}
|\beta_{l+1,i}|_v &\leq |\beta_{l,N-1}a_i|_v + |\beta_{l,i-1}|_v \leq (N+1)|\beta_{l,N-1}|_v|a_i|_v + |\beta_{l,i-1}|_v \\
&\leq (N+1)\exp(c_v l)H_v + \exp(c_v l) = ((N+1)H_v + 1)\exp(c_v l) \\
&= \exp(c_v)\exp(c_v l) = \exp(c_v(l+1)),
\end{aligned}
$$

as desired. If $i = 0$, by using $\beta_{l+1,0} = -\beta_{l,0}a_0$, we can conclude the assertion.

$\square$

Using the above lemma, we show the following proposition:

Proposition 4.29 *Let K be a number field, and let $P(Y,Z)$ and $P'(Y',Z')$ be, respectively, polynomials of two variables Y, Z and Y', Z' over K of the form*

$$
P(Y,Z) = Z^N + a_{N-1}(Y)Z^{N-1} + \cdots + a_0(Y),
$$
$$
P'(Y',Z') = Z'^N + a'_{N-1}(Y')Z'^{N-1} + \cdots + a'_0(Y'),
$$

where $a_i(Y) \in K[Y]$, $\deg(a_i) \leq N - i$ and $a'_i(Y') \in K[Y']$, $\deg(a'_i) \leq N - i$. (So, $\deg(P) = \deg(P') = N$.) Let $X = \mathrm{Spec}(K[Y,Z,Y',Z']/(P,P'))$, and let $\pi: X \to \mathbb{A}^2_K$ be the projection given by $(Y,Z,Y',Z') \mapsto (Y,Y')$. We set

$$
y = Y|_X, \; z = Z|_X, \; y' = Y'|_X, \; z' = Z'|_X.
$$

Then we have the following:

1. *π is finite and flat.*
2. *Let $F \in K[Y,Z,Y',Z']$ be a polynomial of bidegree at most (d,d'), i.e., the degree of F with respect to Y, Z is at most d and the degree of F with respect to Y', Z' is at most d'. Then there are unique $F_{ii'}(Y,Y') \in K[Y,Y']$ $(0 \leq i < N, 0 \leq i' < N)$ such that*

$$
F(y,z,y',z') = \sum_{\substack{0 \leq i < N \\ 0 \leq i' < N}} F_{ii'}(y,y')z^i z'^{i'},
$$

where $F_{ii'}$ is a polynomial of bidegree at most $(d-i, d'-i')$, i.e., the degree of $F_{ii'}$ with respect to Y is at most $d-i$ and the degree of $F_{ii'}$ with respect to Y' is at most $d'-i'$. Further, if all the coefficients of $a_0, \ldots, a_{N-1}, a'_0, \ldots, a'_{N-1}$ and F belong to O_K, then so do all the coefficients of $F_{ii'}$.

3. *For each $v \in M_K$, there is a nonnegative constant c_v depending only on $N, a_0, \ldots, a_{N-1}, a_0', \ldots, a_{N-1}'$ and v such that, for any polynomial $F \in K[Y, Z, Y', Z']$ of bidegree at most (d, d') and for all i, i' $(0 \le i < N, 0 \le i' < N)$, we have*

$$|F_{ii'}|_v \le |F|_v \exp(c_v(d + d')).$$

We can take $c_v = 0$ except for finitely many $v \in M_K$.

4. *For each $v \in M_K$, there is a nonnegative constant c_v' depending only on $N, a_0, \ldots, a_{N-1}, a_0', \ldots, a_{N-1}'$ and v such that, for any polynomial $F \in K[Y, Z, Y', Z']$ of bidegree at most (d, d'), we have*

$$|\operatorname{Norm}_\pi (F(y, z, y', z'))|_v \le |F|_v^{N^2} \exp(c_v'(d + d' + 1)).$$

Further, we can take $c_v' = 0$ except for finitely many v.

Proof 1. Since $K[Y, Z, Y', Z'] / (P(Y, Z), P'(Y', Z'))$ has a free basis $\left\{z^i z'^{i'}\right\}_{\substack{0 \le i < N \\ 0 \le i' < N}}$ as a $K[y, y']$-module, π is finite and flat.

2. The restriction of the natural surjective homomorphism

$$K[Y, Z, Y', Z'] \to K[Y, Z, Y', Z']/(P(Y, Z), P(Y', Z'))$$

to $K[Y, Y']$, i.e.,

$$K[Y, Y'] \to K[Y, Z, Y', Z']/(P(Y, Z), P'(Y', Z')),$$

is injective, so the uniqueness of $F_{ii'}(Y, Y') \in K[Y, Y']$ follows. In the following, let us look at the existence of $F_{ii'}(Y, Y')$ and the other properties.

By Lemma 4.28, for any l and l', there are unique $\beta_{l,i}(Y) \in K[Y]$ and $\beta_{l',i'}'(Y') \in K[Y']$ such that

$$z^l = \sum_{0 \le i < N} \beta_{l,i}(y) z^i, \qquad z'^{l'} = \sum_{0 \le i' < N} \beta_{l',i'}'(y') z'^{i'}$$

and that $\deg(\beta_{l,i}) \le l - i$ and $\deg(\beta_{l',i'}') \le l' - i'$.
We write

$$F = \sum_{\substack{0 \le l \le d \\ 0 \le l' \le d'}} G_{l,l'}(Y, Y') Z^l Z'^{l'}.$$

Then

$$F(y,z,y',z') = \sum_{l,l'} G_{l,l'}(y,y')z^l z'^{l'}$$

$$= \sum_{l,l'} G_{l,l'}(y,y') \left(\sum_{0 \le i < N} \beta_{l,i}(y)z^i \right) \left(\sum_{0 \le i < N} \beta'_{l',i'}(y')z'^{i'} \right)$$

$$= \sum_{\substack{0 \le i < N \\ 0 \le i' < N}} \left(\sum_{l,l'} G_{l,l'}(y,y')\beta_{l,i}(y)\beta'_{l',i'}(y') \right) z^i z'^{i'}.$$

Thus, we have

$$F_{ii'}(Y,Y') = \sum_{l,l'} G_{l,l'}(Y,Y')\beta_{l,i}(Y)\beta'_{l',i'}(Y'). \tag{4.17}$$

Since the bidegrees of $G_{l,l'}$, $\beta_{l,i}$, and $\beta'_{l',i'}$ are at most $(d-l,d'-l')$, $(l-i,0)$, and $(0,l'-i')$, respectively, the bidegree of $F_{ii'}(Y,Y')$ is at most $(d-i,d'-i')$.

Finally, we assume that $a_0,\ldots,a_{N-1} \in O_K[Y]$ and $a'_0,\ldots,a'_{N-1} \in O_K[Y']$. Then, by Lemma 4.28, $\beta_{l,i} \in O_K[Y]$ and $\beta'_{l',i'} \in O_K[Y']$. Further, if we assume that all the coefficients of F belong to O_K, i.e., $G_{l,l'} \in O_K[Y,Y']$ for any l,l', then, by (4.17), all the coefficients of $F_{ii'}$ belong to O_K.

3. We keep the notation in the proof of 2. Lemma 4.28 tells us that there is a nonnegative constant c_v depending only on $N, a_0,\ldots,a_{N-1}$, $a'_0,\ldots,a'_{N-1}$ and v such that, for any l,i,l',i', $|\beta_{l,i}|_v \le \exp(c_v l)$ and $|\beta'_{l',i'}|_v \le \exp(c_v l')$. Further, we can take $c_v = 0$ except for finitely many v.

We evaluate the length of $F_{ii'}$ with respect to v. First, suppose that v is nonarchimedean. Then, Gauss's lemma (Proposition 4.4), (4.3), and Lemma 4.28 tell us that

$$|F_{ii'}|_v \le \max_{l,l'}\{|G_{l,l'}\beta_{l,i}\beta'_{l',i'}|_v\} = \max_{l,l'}\{|G_{l,l'}|_v|\beta_{l,i}|_v|\beta'_{l',i'}|_v\}$$

$$\le |F|_v \exp(c_v(d+d')).$$

Next, suppose that v is archimedean. Note that $l \in \{0,\ldots,d\}$ and $l' \in \{0,\ldots,d'\}$ in Equation (4.17). Further, the bidegree of $\beta_{l,i}$ and $\beta'_{l',i'}$ are at most $(d,0)$ and $(0,d')$, respectively. It follows that

$$\begin{cases} (1 + \deg_Y(\beta_{l,i}))(1 + \deg_{Y'}(\beta_{l,i})) \le 1 + d, \\ (1 + \deg_Y(\beta'_{l',i'}))(1 + \deg_{Y'}(\beta'_{l',i'})) \le 1 + d'. \end{cases}$$

Then, by (4.3),

$$|\beta_{l,i}\beta'_{l',i'}G_{l,l'}|_v \le |\beta_{l,i}|_v|\beta'_{l',i'}|_v|G_{l,l'}|_v(1+d)(1+d').$$

Combining Corollary 4.6, (4.4), and Lemma 4.28, we have

$$|F_{ii'}|_v \leq \sum_{l,l'} |G_{l,l'}\beta_{l,i}\beta'_{l',i'}|_v$$

$$\leq \sum_{l,l'} (1+d)(1+d')|G_{l,l'}|_v|\beta_{l,i}|_v|\beta'_{l',i'}|_v$$

$$\leq \sum_{l,l'} (1+d)(1+d')|F|_v \exp(c_v(d+d'))$$

$$\leq (1+d)^2(1+d')^2|F|_v \exp(c_v(d+d'))$$

$$\leq \exp(2d)\exp(2d')|F|_v \exp(c_v(d+d')) = |F|_v \exp((c_v+2)(d+d')).$$

Here, for the last equality, we use the inequality $\exp(t) \geq 1+t$ ($t \geq 0$). Thus, when v is archimedean, replacing c_v by $c_v + 2$, we obtain the assertion.

4. For $0 \leq j < N, 0 \leq j' < N$, we set $F^{jj'} = Z^j Z'^{j'} F$ and

$$F^{jj'}(y,z,y',z') = \sum_{\substack{0 \leq i < N \\ 0 \leq i' < N}} F^{(j,j')}_{(i,i')}(y,y')z^i z'^{i'}.$$

By 2 and 3, $F^{(j,j')}_{(i,i')}$ is a polynomial of bidegree at most $(d+j-i, d'+j'-i')$ and

$$\left| F^{(j,j')}_{(i,i')} \right|_v \leq |F^{jj'}|_v \exp(c_v(d+d'+j+j')))$$

$$= |F|_v \exp(c_v(d+d'+j+j')).$$

In particular, $F^{(j,j')}_{(i,i')}$ is a polynomial of bidegree at most $(d+N-1, d'+N-1)$ and

$$\left| F^{(j,j')}_{(i,i')} \right|_v \leq |F|_v \exp(c_v(d+d'+2(N-1))). \tag{4.18}$$

Let A_F be the matrix representation of the homomorphism given by

$$K[y,z,y'z'] \to K[y,z,y'z'] \quad (\psi \mapsto F\psi)$$

with respect to the basis $\{z^j z'^{j'} \mid 0 \leq j, j' < N\}$ as a $K[y,y']$-module. Then $\mathrm{Norm}(F) = \det(A_F)$ and the entries of the matrix A_F consist of $F^{(j,j')}_{(i,i')}$. Let $\mathfrak{S}$ be the permutation group on $S = \{(j,j') \in \mathbb{Z}^2 \mid 0 \leq j, j' < N\}$. For $\sigma \in \mathfrak{S}$, we set

$$G_\sigma := \prod_{(j,j') \in S} F^{(j,j')}_{\sigma(j,j')}.$$

First, suppose that v is nonarchimedean. Then Gauss's lemma (Proposition 4.4) and (4.18) tell us that

$$|G_\sigma|_v \le |F|_v^{N^2} \exp(c_v N^2(d + d' + 2(N - 1))).$$

Thus,

$$|\operatorname{Norm}(F)|_v \le |F|_v^{N^2} \exp(c_v N^2(d + d' + 2(N - 1))).$$

Putting $c_v' = c_v N^2(2(N - 1))$, we obtain the assertion.

Next, suppose that v is archimedean. Note that $F_{(i,i')}^{(j,j')}$ is a polynomial of two variables Y and Y', whose bidegree is at most $(d + N - 1, d' + N - 1)$. It follows that

$$\left(1 + \deg_Y \left(F_{(i,i')}^{(j,j')}\right)\right) \left(1 + \deg_{Y'} \left(F_{(i,i')}^{(j,j')}\right)\right) \le (d + N)(d' + N).$$

On the other hand, by 2,

$$\left|F_{(i,i')}^{(j,j')}\right|_v \le |F^{jj'}|_v \exp(c_v(d + d' + 2(N - 1)))$$
$$= |F|_v \exp(c_v(d + d' + 2(N - 1))),$$

and thus, by Corollary 4.6, we have

$$|G_\sigma|_v \le |F|_v^{N^2} \exp(c_v N^2(d + d' + 2(N - 1))) \left((d + N)(d' + N)\right)^{N^2 - 1}.$$

Since $\exp(t) \ge 1 + t$ for $t \ge 0$, we obtain

$$\left((d + N)(d' + N)\right)^{N^2 - 1} \le \exp\left((N^2 - 1)(d + N - 1)\right) \times$$
$$\exp\left((N^2 - 1)(d' + N - 1)\right) \le \exp\left(N^2(d + d' + 2(N - 1))\right),$$

which implies

$$|G_\sigma|_v \le |F|_v^{N^2} \exp\left((c_v + 1)N^2(d + d' + 2(N - 1))\right).$$

It follows that

$$|\operatorname{Norm}(F)|_v \le |F|_v^{N^2} \exp((c_v + 1)N^2(d + d' + 2(N - 1)))(N^2)!.$$

Since $\log(t) \le t - 1$ for $t \ge 1$, we have

$$(N^2)! \le \exp(N^2 \log(N^2)) \le \exp(N^2 \cdot 2(N - 1)) \le \exp(N^2(d + d' + 2(N - 1))).$$

Thus,

$$|\operatorname{Norm}(F)|_v \le |F|_v^{N^2} \exp((c_v + 2)N^2(d + d' + 2(N - 1))).$$

Putting $c_v' = (c_v + 2)N^2(2(N - 1))$ when v is archimedean, we obtain the assertion. $\square$

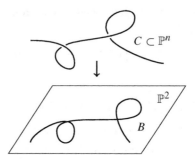

Figure 4.4 Projection from $\mathbb{P}^n$ to $\mathbb{P}^2$.

In the rest of this section, we give a result that will be needed in Chapter 5. First, we fix our setting. Let K be a number field, let V be an $(n+1)$-dimensional K-vector space ($n \geq 2$), let $\mathbb{P}(V)$ be the associated projective space, and let C be a geometrically irreducible, smooth projective curve in $\mathbb{P}(V)$ over K such that the degree of C in $\mathbb{P}(V)$ is N.

We take a basis of V, and let $(X_0 : X_1 : \ldots : X_n)$ denote the homogeneous coordinates of $\mathbb{P}(V)$ associated to it. We also take another basis of V, and let $(Y_0 : Y_1 : \ldots : Y_n)$ denote the homogeneous coordinates of $\mathbb{P}(V)$ associated to it.[4]

We assume that

$$C \cap \{X_0 = X_1 = 0\} = C \cap \{Y_0 = Y_1 = 0\} = \emptyset \tag{4.19}$$

Let B and B' be the images of the projections (Figure 4.4) given by $(X_0 : \ldots : X_n) \mapsto (X_0 : X_1 : X_2)$ and $(Y_0 : \ldots : Y_n) \mapsto (Y_0 : Y_1 : Y_2)$, respectively. We assume that C is birational to B and B' by the projections, respectively. Then $\deg(B) = \deg(B') = N$. These assumptions are satisfied if we choose general bases of V. Note that B and B' do not pass through the point $(0 : 0 : 1)$ by (4.19), so B and B' are defined by the following homogeneous polynomials of degree N:

$$\begin{cases} B : X_2^N + a_{N-1}(X_0, X_1)X_2^{N-1} + \cdots + a_1(X_0, X_1)X_2 + a_0(X_0, X_1), \\ B' : Y_2^N + a'_{N-1}(Y_0, Y_1)Y_2^{N-1} + \cdots + a'_1(Y_0, Y_1)Y_2 + a'_0(Y_0, Y_1). \end{cases}$$

---

[4] Proposition 4.30 will be used in Section 5.4, where we consider the projections of C given by the different directions $(X_0 : X_1 : \ldots : X_n) \mapsto (X_0 : X_i)$ and $(X_0 : X_1 : \ldots : X_n) \mapsto (X_0 : X_j)$. Here, to avoid complications of the notation, we introduce two different coordinates $(X_0 : X_1 : \ldots : X_n)$ and $(Y_0 : Y_1 : \ldots : Y_n)$ and use the projections $(X_0 : X_1 : \ldots : X_n) \mapsto (X_0 : X_1)$ and $(Y_0 : Y_1 : \ldots : Y_n) \mapsto (Y_0 : Y_1)$. Thus, in Section 5.4, the second basis of V we consider is a permutation of the first basis of V we have chosen.

Let $f : C \to \mathbb{P}^1$ and $f' : C \to \mathbb{P}^1$ be the morphisms given by the projections $(X_0 : \ldots : X_n) \mapsto (X_0 : X_1)$ and $(Y_0 : \ldots : Y_n) \mapsto (Y_0 : Y_1)$, respectively, and we denote $f \times f' : C \times C \to \mathbb{P}^1 \times \mathbb{P}^1$ by π. Note that f and f' are finite and flat, so π is also finite and flat. Further, let $p_i : C \times C \to C$ and $q_i : \mathbb{P}^1 \times \mathbb{P}^1 \to \mathbb{P}^1$ be the projections to the i-th factor.

Note that $(X_0 : \ldots : X_n)$ and $(Y_0 : \ldots : Y_n)$ are homogeneous coordinates of the first factor and the second factor of $C \times C$ ($\subset \mathbb{P}(V) \times \mathbb{P}(V)$), respectively. We set $x_i = X_i|_C$ and $y_i = Y_i|_C$ for $i = 0, \ldots, n$.

We identify $\mathbb{P}(V) \times \mathbb{P}(V)$ with $\mathbb{P}^n \times \mathbb{P}^n$ by fixing homogeneous coordinates $(X_0 : \ldots : X_n)$ and $(Y_0 : \ldots : Y_n)$. The vector space consisting of bihomogeneous polynomials F of bidegree (d, d') in

$$K[X_0, \ldots, X_n, Y_0, \ldots, Y_n]$$

is isomorphic to the space of global sections of the invertible sheaf $\mathcal{O}_{\mathbb{P}^n \times \mathbb{P}^n}(d, d')$ on $\mathbb{P}^n \times \mathbb{P}^n$. Via $C \times C \subset \mathbb{P}(V) \times \mathbb{P}(V) \cong \mathbb{P}^n \times \mathbb{P}^n$, we denote the restriction of $\mathcal{O}_{\mathbb{P}^n \times \mathbb{P}^n}(d, d')$ to $C \times C$ by $\mathcal{O}_{C \times C}(d, d')$. Then $F(x_0, \ldots, x_n, y_0, \ldots, y_n)$ gives an element of $H^0(C \times C, \mathcal{O}_{C \times C}(d, d'))$. Since π is flat, by the definition of the norm in Section 4.5, we have

$$\operatorname{Norm}_{\pi}(F(x_0, \ldots, x_n, y_0, \ldots, y_n)) \in H^0(\mathbb{P}^1 \times \mathbb{P}^1, \operatorname{Norm}_{\pi}(\mathcal{O}_{C \times C}(d, d'))).$$

In the following, we will determine $\operatorname{Norm}_{\pi}(\mathcal{O}_{C \times C}(d, d'))$ and evaluate the height of $\operatorname{Norm}_{\pi}(F(x_0, \ldots, x_n, y_0, \ldots, y_n))$.

Proposition 4.30 *1.* $\operatorname{Norm}_{\pi}(\mathcal{O}_{C \times C}(d, d')) = \mathcal{O}_{\mathbb{P}^1 \times \mathbb{P}^1}(N^2 d, N^2 d')$.
2. Let $F \in K[X_0, \ldots, X_n, Y_0, \ldots, Y_n] \setminus \{0\}$ be a bihomogeneous polynomial of bidegree (d, d'). Then, by 1,

$$\operatorname{Norm}_{\pi}(F(x_0, \ldots, x_n, y_0, \ldots, y_n))$$

is a bihomogeneous polynomial of bidegree $(N^2 d, N^2 d')$ in

$$K[x_0, x_1, y_0, y_1].$$

Further, there is a positive constant M depending only on C, $a_0, \ldots, a_{N-1}$, and $a'_0, \ldots, a'_{N-1}$ such that

$$h(\operatorname{Norm}_{\pi}(F(x_0, \ldots, x_n, y_0, \ldots, y_n))) \leq N^2 h(F) + M(d + d' + 1).$$

Proof 1. For $i = 0, 1$, $j = 0, 1$, let $U_{(i,j)}$ be the Zariski open set on $\mathbb{P}^1 \times \mathbb{P}^1$ given by $\{X_i \neq 0, Y_j \neq 0\}$. Then $x_i^d y_j^{d'}$ gives a local basis of $\mathcal{O}_{C \times C}(d, d')$ over $\pi^{-1}(U_{(i,j)})$. Thus, if we set

$$g_{(i,j)(i',j')} = \frac{x_i^d y_j^{d'}}{x_{i'}^d y_{j'}^{d'}} = (x_i/x_{i'})^d (y_j/y_{j'})^{d'}$$

over $\pi^{-1}(U_{(i,j)} \cap U_{(i',j')})$, then $\mathcal{O}_{C \times C}(d, d')$ is given by $\{g_{(i,j)(i',j')}\}$. Since

$$\mathrm{Norm}_\pi(g_{(i,j)(i',j')}) = (x_i/x_{i'})^{N^2 d}(y_j/y_{j'})^{N^2 d'},$$

$\mathrm{Norm}_\pi(\mathcal{O}_{C \times C}(d, d'))$ is given by $\left\{(x_i/x_{i'})^{N^2 d}(y_j/y_{j'})^{N^2 d'}\right\}$. We have shown 1.

2. Since

$$h\left(\mathrm{Norm}_\pi\left(F(1, x_1/x_0, \ldots, x_n/x_0, \ldots, 1, y_1/y_0, \ldots, y_n/y_0)\right)\right)$$
$$= h\left(\mathrm{Norm}_\pi\left(\frac{F}{x_0^d y_0^{d'}}\right)\right) = h\left(\frac{\mathrm{Norm}_\pi(F)}{x_0^{N^2 d} y_0^{N^2 d'}}\right) = h\left(\mathrm{Norm}_\pi(F)\right),$$

it suffices to consider the evaluation of the height on $\pi^{-1}(U_{(0,0)})$. Thus, we may replace X_i/X_0 by X_i, Y_i/Y_0 by Y_i, x_i/x_0 by x_i, and y_i/y_0 by y_i for $i = 1, \ldots, n$. Then B and B' are given by the equations

$$X_2^N + a_{N-1}(1, X_1)X_2^{N-1} + \cdots + a_1(1, X_1)X_2 + a_0(1, X_1),$$
$$Y_2^N + a'_{N-1}(1, Y_1)Y_2^{N-1} + \cdots + a'_1(1, Y_1)Y_2 + a'_0(1, Y_1),$$

respectively.

Since C is birational to B, the support of $K[x_1, \ldots, x_n]/K[x_1, x_2]$ as a $K[x_1, x_2]$-module is finite. Further, $K[x_1, x_2]$ is finitely generated as a $K[x_1]$-module, so there is a nonzero polynomial $\gamma(S) \in K[S]$ such that

$$\gamma(x_1)K[x_1, \ldots, x_n] \subseteq K[x_1, x_2].^5$$

In particular, for any $i \geq 3$, there is $g_i(U, V) \in K[U, V]$ such that $\gamma(x_1)x_i = g_i(x_1, x_2)$. In the same way, there are a nonzero polynomial $\gamma'(S) \in K[S]$ and nonzero polynomials $g'_i \in K[U, V]$ $(i = 3, \ldots, n)$ such that $\gamma'(y_1)y_i = g'_i(y_1, y_2)$ for $i = 3, \ldots, n$.

---

[5] This follows from the following fact: Let A and B be Noetherian domains such that $A \subseteq B$ and B is a finitely generated over A as an A-module. If M is a finitely generated B-module such that $bM = \{0\}$ for some $b \in B \setminus \{0\}$, then there is $a \in A \setminus \{0\}$ such that $aM = \{0\}$. Indeed, since B is integral over A, there are $n \geq 1$ and $a_1, \ldots, a_n \in A$ such that $b^n + a_1 b^{n-1} + \cdots + a_n = 0$ and $a_n \neq 0$. If we set $b' = b^{n-1} + a_1 b^{n-2} + \cdots + a_{n-1} \in B$, then $b'b = -a_n \in A \setminus \{0\}$, as desired.

We write

$$F(1, X_1, \ldots, X_n, 1, Y_1, \ldots, Y_n)$$
$$= \sum_{i_3,\ldots,i_n,i_3',\ldots,i_n'} F_{i_3,\ldots,i_n,i_3',\ldots,i_n'}(X_1, X_2, Y_1, Y_2) X_3^{i_3} \cdots X_n^{i_n} Y_3^{i_3'} \cdots Y_n^{i_n'}, \quad (4.20)$$

where $F_{i_3,\ldots,i_n,i_3',\ldots,i_n'}$ is a polynomial of bidegree at most $(d - i_3 - \cdots - i_n, d' - i_3' - \cdots - i_n')$. Then

$$\gamma(X_1)^d \gamma'(Y_1)^{d'} F(1, X_1, \ldots, X_n, 1, Y_1, \ldots, Y_n)$$
$$= \sum \gamma(X_1)^{d-(i_3+\cdots+i_n)} \gamma'(Y_1)^{d'-(i_3'+\cdots+i_n')} F_{i_3,\ldots,i_n,i_3',\ldots,i_n'} \times$$
$$(\gamma(X_1)X_3)^{i_3} \cdots (\gamma(X_1)X_n)^{i_n} (\gamma'(Y_1)Y_3)^{i_3'} \cdots (\gamma'(Y_1)Y_n)^{i_n'}.$$

Let us consider the evaluation $|T|_v$ of

$$T = \sum_{i_3,\ldots,i_n,i_3',\ldots,i_n'} \gamma(X_1)^{d-(i_3+\cdots+i_n)} \gamma'(Y_1)^{d'-(i_3'+\cdots+i_n')} F_{i_3,\ldots,i_n,i_3',\ldots,i_n'} \times$$
$$g_3(X_1, X_2)^{i_3} \cdots g_n(X_1, X_2)^{i_n} g_3'(Y_1, Y_2)^{i_3'} \cdots g_n'(Y_1, Y_2)^{i_n'}$$

at each place v. Note that $d \geq i_3 + \cdots + i_n$ and $d' \geq i_3' + \cdots + i_n'$ by (4.20). We set

$$\begin{cases} A_v = \max\{1, |\gamma|_v, |\gamma'|_v, |g_3|_v, |g_3'|_v, \ldots, |g_n|_v, |g_n'|_v\}, \\ s = \max\{\deg(\gamma), \deg(\gamma'), \deg(g_3), \deg(g_3'), \ldots, \deg(g_n), \deg(g_n')\}. \end{cases}$$

Note that $A_v = 1$ except for finitely many v.

Suppose that v is nonarchimedean. Then Gauss's lemma (Proposition 4.4) and (4.3) tell us that

$$|T|_v \leq \max\left\{ A_v^{d+d'-(i_3+\cdots+i_n+i_3'+\cdots+i_n')} |F|_v A_v^{i_3} \cdots A_v^{i_n} \cdot A_v^{i_3'} \cdots A_v^{i_n'} \right\}$$
$$= A_v^{d+d'} |F|_v.$$

Suppose that v is archimedean. The evaluation will become a little complicated. First, we consider

$$\Delta_{i_3,\ldots,i_n,i_3',\ldots,i_n'} = \gamma(X_1)^{d-(i_3+\cdots+i_n)} \gamma'(Y_1)^{d'-(i_3'+\cdots+i_n')} F_{i_3,\ldots,i_n,i_3',\ldots,i_n'} \times$$
$$g_3(X_1, X_2)^{i_3} \cdots g_n(X_1, X_2)^{i_n} g_3'(Y_1, Y_2)^{i_3'} \cdots g_n'(Y_1, Y_2)^{i_n'}.$$

We claim that $|\Delta_{i_3,\ldots,i_n,i_3',\ldots,i_n'}|_v$ is bounded by $(1+s)^{2(d+d')} A_v^{d+d'} |F|_v$. Indeed, since

$$\begin{cases} 1 + \deg_{X_1}(\gamma(X_1)) \le 1 + s \le (1+s)^2, \\ 1 + \deg_{X_2}(\gamma(X_2)) \le 1 + s \le (1+s)^2, \\ (1 + \deg_{X_1}(g_3(X_1, X_2)))(1 + \deg_{X_2}(g_3(X_1, X_2))) \le (1+s)^2, \end{cases}$$

Corollary 4.6 tells us that

$$\left| \Delta_{i_3, \dots, i_n, i'_3, \dots, i'_n} \right|_v \le A_v^{d+d'} (1+s)^{2(d+d')} \left| F_{i_3, \dots, i_n, i'_3, \dots, i'_n} \right|_v$$

$$\le A_v^{d+d'} (1+s)^{2(d+d')} |F|_v.$$

Since $d \ge i_3 + \cdots + i_n$ and $d' \ge i'_3 + \cdots + i'_n$, we have $i_3, \dots, i_n \in \{0, \dots, d\}$ and $i'_3, \dots, i'_n \in \{0, \dots, d'\}$. Thus, using (4.4), we have

$$|T|_v \le \sum_{i_3, \dots, i_n, i'_3, \dots, i'_n} \left| \Delta_{i_3, \dots, i_n, i'_3, \dots, i'_n} \right|_v \le \sum_{i_3, \dots, i_n, i'_3, \dots, i'_n} (1+s)^{2(d+d')} A_v^{d+d'} |F|_v$$

$$\le (1+d)^{n-2}(1+d')^{n-2} \left((1+s)^{2(d+d')} A_v^{d+d'} |F|_v \right).$$

Since $1+d \le 2^d$ and $1+d' \le 2^{d'}$ for nonnegative integers d and d', we obtain

$$|T|_v \le (2^{n-2}(1+s)^2 A_v)^{d+d'} |F|_v.$$

Thus, when v is archimedean, replacing A_v by $2^{n-2}(1+s)^2 A_v$, we have $|T|_v \le A_v^{d+d'} |F|_v$.

Since T is a polynomial of bidegree at most $(d + ds, d' + d's) = ((1+s)d, (1+s)d')$, it follows from Proposition 4.29 that, for each place v, there is a nonnegative constant c'_v depending only on N, $a_0, \dots, a_{N-1}$, $a'_0, \dots, a'_{N-1}$, and v such that $c'_v = 0$ except for finitely many v and

$$|\operatorname{Norm}(T(x_1, x_2, y_1, y_1))|_v \le |T|_v^{N^2} \exp(c'_v((1+s)(d+d') + 1).$$

Thus,

$$|\operatorname{Norm}(T(x_1, x_2, y_1, y_1))|_v \le A_v^{N^2(d+d')} |F|_v^{N^2} \exp(c'_v((1+s)(d+d') + 1)$$

$$\le A_v^{N^2(d+d'+1)} |F|_v^{N^2} \exp(c'_v((1+s)(d+d'+1)).$$

It follows that, if we set $M = \sum_v (N^2 \log(A_v) + (1+s)c'_v)$, then

$$h(\operatorname{Norm}(T(x_1, x_2, y_1, y_1))) \le N^2 h(F) + M(d + d' + 1).$$

On the other hand, since

$$T(x_1, x_2, y_1, y_2) = \gamma(x_1)^d \gamma'(y_1)^{d'} F(1, x_1, \dots, x_n, 1, y_1, \dots, y_n),$$

we obtain

$$\operatorname{Norm}(T(x_1, x_2, y_1, y_1)) = \gamma(x_1)^{N^2 d} \gamma'(y_1)^{N^2 d'} \times$$

$$\operatorname{Norm}(F(1, x_1, \dots, x_n, 1, y_1, \dots, y_n)).$$

Since $\mathrm{Norm}(F(1,x_1,\ldots,x_n,1,y_1,\ldots,y_n))$ is an element of $K[x_1,y_1]$ whose bidegree is at most (N^2d, N^2d'), it follows from Proposition 3.5, 3 and Proposition 4.8, 2 that

$$h(\mathrm{Norm}(F(1,x_1,\ldots,x_n,1,y_1,\ldots,y_n)))$$
$$\leq h(\mathrm{Norm}(F(1,x_1,\ldots,x_n,1,y_1,\ldots,y_n))) + h(\gamma(x_1)^{N^2d}\gamma'(y_1)^{N^2d'})$$
$$\leq h(\mathrm{Norm}(T(x_1,x_2,y_1,y_1)))$$
$$\qquad + 2(N^2(d+d')\max\{\deg(\gamma),\deg(\gamma')\} + N^2(d+d'))$$
$$\leq N^2 h(F) + M(d+d'+1)$$
$$\qquad + 2(N^2(d+d')\max\{\deg(\gamma),\deg(\gamma')\} + N^2(d+d')).$$

Now, replacing M by $M + 2N^2(\max\{\deg(\gamma),\deg(\gamma')\} + 1)$, we have

$$h(\mathrm{Norm}(F(x_1,\ldots,x_n,y_1,\ldots,y_n))) \leq N^2 h(F) + M(d+d'+1),$$

as desired. $\qquad\qquad\qquad\qquad\qquad\qquad\qquad\qquad\qquad\qquad\qquad\qquad\square$

4.7 Local Eisenstein Theorem

In this section, we prove the local Eisenstein theorem. Let K be a number field and let C be a geometrically irreducible, smooth projective curve over K. Let $\pi : C \to \mathbb{P}^1$ be a surjective morphism over K (Figure 4.5). We define a finite subset S of $\mathbb{P}^1$ by

$$S = \{a \in \mathbb{P}^1 \mid \pi \text{ is not étale over } a\} \cup \{\infty\}.$$

Further, we set $C_0 = C \setminus \pi^{-1}(S)$. We write T for a coordinate of $\mathbb{P}^1 \setminus \{\infty\} = \mathbb{A}^1$, i.e., $\mathbb{P}^1 \setminus \{\infty\} = \mathrm{Spec}(K[T])$. For $z \in C_0(K)$, $x = T - a$ ($a = \pi(x) \in K$) gives a local coordinate of C around z. If y is a rational

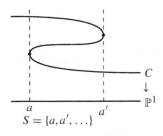

Figure 4.5 Ramified covering from C to $\mathbb{P}^1$.

Figure 4.6 Gotthold Eisenstein.
Source: Archives of the Mathematisches Forschungsinstitut Oberwolfach.

function on C defined over K, then y is algebraic over $K(T)$, so there is a polynomial $P(X, Y) \in K[X, Y]$ such that $P(x, y) = 0$. Further, as z is a K-rational point of C, the completion $\widehat{\mathcal{O}}_{C,z}$ of $\mathcal{O}_{C,z}$ is isomorphic to $K[[x]]$ (for example, see [18, Theorem 29.7] or Footnote 2 on p. 85), so the coefficients of the Taylor expansion of an element of $\mathcal{O}_{C,z}$ at z belong to K.

Thus, if y has no pole at z, then, for all $l \geq 0$,

$$\left(\left(\frac{d}{dx} \right)^{l} y \right)(0) \in K,$$

where the symbol (0) in the above formula means the evaluation at $x = 0$. Then we have the following local Eisenstein theorem (Figure 4.6):

Lemma 4.31 (Local Eisenstein theorem)   *We assume that y has no pole at z and $(\partial P / \partial Y)(0, y(0)) \neq 0$. Let v be a place of K such that $|y(0)|_v \leq 1$. Let $|P|_v$ denote the length of the polynomial P with respect to v (see Section 4.2). Then, for any $l \geq 0$, we have the following:*

1. If v is nonarchimedean, then

$$\left| \frac{1}{l!} \left(\left(\frac{d}{dx} \right)^{l} y \right)(0) \right|_v \leq \left(\frac{|P|_v}{|(\partial P / \partial Y)(0, y(0))|_v} \right)^{\max\{2l-1, 0\}}.$$

2. If v is archimedean, then

$$\left| \frac{1}{l!} \left(\left(\frac{d}{dx} \right)^{l} y \right)(0) \right|_v \leq (2 \deg(P))^{7l} \left(\frac{|P|_v}{|(\partial P / \partial Y)(0, y(0))|_v} \right)^{\max\{2l-1, 0\}}.$$

Proof To simplify the notation, we denote the l-th derivative of y with respect to x by $y^{(l)}$. For a polynomial $F(X, Y) \in K[X, Y]$, we denote the partial derivative $\partial^{i+j} F / \partial X^i \partial Y^j$ of F by $F_{X^i Y^j}$.

We define a sequence $\{Q_l\}_{l=1}^{\infty}$ by the following recursive equation:

$$\begin{cases} Q_1 = P_X, \\ Q_{l+1} = (Q_l)_X (P_Y)^2 - (Q_l)_Y P_X P_Y + (2l - 1) Q_l (P_{YY} P_X - P_{XY} P_Y). \end{cases}$$
(4.21)

We claim that

$$Q_l(x, y) + (P_Y(x, y))^{2l-1} y^{(l)} = 0 \tag{4.22}$$

for all $l \geq 1$. We show (4.22) by induction on l. Indeed, considering the derivative of $P(x, y) = 0$ with respect to x, we have

$$P_X(x, y) + P_Y(x, y) y' = 0. \tag{4.23}$$

This is the case $l = 1$ of (4.22). We assume that (4.22) holds for $l \geq 1$. Taking the derivative of (4.22) with respect to x, we obtain

$$(Q_l)_X(x, y) + (Q_l)_Y(x, y) y'$$
$$+ (2l - 1)(P_Y(x, y))^{2l-2}(P_{XY}(x, y) + P_{YY}(x, y) y') y^{(l)}$$
$$+ (P_Y(x, y))^{2l-1} y^{(l+1)} = 0.$$

Note that $P_X(x, y) + P_Y(x, y) y' = 0$ and $Q_l(x, y) + (P_Y(x, y))^{2l-1} y^{(l)} = 0$. Thus, multiplying $P_Y(x, y)^2$ to the above equation, we have

$$Q_{l+1}(x, y) + (P_Y(x, y))^{2l+1} y^{(l+1)} = 0,$$

as required.

We set $n = \deg(P)$. Equation (4.21) tells us that Q_l is a polynomial of degree at most $(2l - 1)(n - 1)$.

1. Suppose that v is archimedean. If $l = 0$, then $|y(0)|_v \leq 1$ by the assumption. Thus, we may assume that $l \geq 1$. We observe

$$|P_X|_v \leq n |P|_v, \quad |P_Y|_v \leq n |P|_v, \quad |P_{XY}|_v \leq n^2 |P|_v.$$

It follows from the recursive equation (4.21), Corollary 4.6, and (4.4) that

$$|Q_{l+1}|_v \leq |(Q_l)_X P_Y^2|_v + |(Q_l)_Y P_X P_Y|_v$$
$$+ (2l - 1)|Q_l P_{YY} P_X|_v + (2l - 1)|Q_l P_{XY} P_Y|_v$$
$$\leq n^4 (|(Q_l)_X|_v |P_Y|_v^2 + |(Q_l)_Y|_v |P_X|_v |P_Y|_v$$
$$+ (2l - 1)|Q_l|_v |P_{YY}|_v |P_X|_v + (2l - 1)|Q_l|_v |P_{XY}|_v |P_Y|_v)$$
$$\leq n^4 \{4(2l - 1)n^3\} |P|_v^2 |Q_l|_v \leq 8n^7 l |P|_v^2 |Q_l|_v.$$

Thus,

$$|Q_l|_v \leq (8n^7)^{l-1}(l-1)!\,|Q_1|_v|P|_v^{2(l-1)} \leq n(8n^7)^{l-1}(l-1)!\,|P|_v^{2l-1}.$$

Note that $Q_l(0, y)$ has the nonzero terms at most $(2l - 1)(n - 1) + 1$ and $|y(0)|_v \leq 1$, so

$$|Q_l(0, y(0))|_v \leq ((2l - 1)(n - 1) + 1)|Q_l|_v$$
$$\leq (2ln)n(8n^7)^{l-1}(l-1)!\,|P|_v^{2l-1} \leq l!\,(2n)^{7l}|P|_v^{2l-1}.$$

Thus, by (4.22), we obtain the assertion, because

$$|Q_l(0, y(0))|_v = |P_Y(0, y(0))|_v^{2l-1}|y^{(l)}(0)|_v.$$

2. Suppose that v is nonarchimedean. Since similar computation as in 1 does not work well, we will compute differently. For $l \geq 0$, we set $\partial_l = \dfrac{1}{l!}\dfrac{d^l}{dx^l}$. Note that the notation ∂_l makes the Leibniz rule simpler, i.e.,

$$\partial_l(f_1 \cdots f_r) = \sum_{a_1 + \cdots + a_r = l} \partial_{a_1}(f_1) \cdots \partial_{a_r}(f_r).$$

We write $P(X, Y) = \sum_{ij} p_{ij} X^i Y^j$. We prove the assertion by induction on l. If $l = 0$, then $|y(0)|_v \leq 1$ by the assumption. If $l = 1$, then (4.23) gives

$$|P_Y(0, y(0))|_v|y'(0)|_v = |P_X(0, y(0))|_v \leq |P|_v,$$

as desired. Thus, we may assume that $l \geq 2$. By the Leibniz rule,

$$0 = \partial_l(P(x, y))$$
$$= \sum_{ij} p_{ij} \sum_{e_0 + e_1 + \cdots + e_j = l} \partial_{e_0}(x^i)\partial_{e_1}(y) \cdots \partial_{e_j}(y)$$
$$= \sum_{ij} p_{ij} j x^i y^{j-1} \partial_l(y) + \sum_{ij} p_{ij} \sum_{\substack{e_0 + e_1 + \cdots + e_j = l \\ 0 \leq e_1 < l, \ldots, 0 \leq e_j < l}} \partial_{e_0}(x^i)\partial_{e_1}(y) \cdots \partial_{e_j}(y)$$
$$= P_Y(x, y)\partial_l(y) + \sum_{ij} p_{ij} \sum_{\substack{e_0 + e_1 + \cdots + e_j = l \\ 0 \leq e_1 < l, \ldots, 0 \leq e_j < l}} \partial_{e_0}(x^i)\partial_{e_1}(y) \cdots \partial_{e_j}(y).$$

Since

$$\partial_{e_0}(x^i)(0) = \begin{cases} 1 & (\text{if } e_0 = i), \\ 0 & (\text{if } e_0 \neq i), \end{cases}$$

we have

$$|P_Y(0, y(0))\partial_l(y)(0)|_v \leq |P|_v \max_{\substack{e_1 + \cdots + e_j = l - i, \\ 0 \leq i \leq l, \\ 0 \leq e_1 < l, \ldots, 0 \leq e_j < l}} \{|\partial_{e_1}(y)(0)|_v \cdots |\partial_{e_j}(y)(0)|_v\}.$$

On the other hand, under the assumptions $e_1 + \cdots + e_j = l - i, 0 \leq i \leq l$ and $0 \leq e_1 < l, \ldots, 0 \leq e_j < l$, we have

$$\max\{2e_1 - 1, 0\} + \cdots + \max\{2e_j - 1, 0\} \leq 2l - 2.$$

Thus, noting that $|P_Y(0, y(0))|_v \leq |P_Y|_v \leq |P|_v$, we have by the induction hypothesis that

$$|P_Y(0, y(0))|_v |\partial_l(y)(0)|_v = |P_Y(0, y(0))\partial_l(y)(0)|_v$$

$$\leq |P|_v \left(\frac{|P|_v}{|P_Y(0, y(0))|_v} \right)^{\max\{2e_1 - 1, 0\} + \cdots + \max\{2e_j - 1, 0\}}$$

$$\leq |P|_v \left(\frac{|P|_v}{|P_Y(0, y(0))|_v} \right)^{2l - 2}.$$

Thus, dividing the above inequalities by $|P_Y(0, y(0))|_v$, we obtain the assertion when v is nonarchimedean. $\square$

In the rest of this section, we give a multiple version of Lemma 4.31, which is useful in Chapter 5. Under the same setting of Lemma 4.31, let $y_1, \ldots, y_e$ be rational functions on C defined over K and we choose $P_i(X, Y) \in K[X, Y]$ such that $P_i(x, y_i) = 0$. Further, as before, we set $\partial_l = \frac{1}{l!} \frac{d^l}{dx^l}$. Then we have the following corollary:

Corollary 4.32 *Let v be a place of K. We assume that $y_1, \ldots, y_e$ have no pole at z and, for all i, $|y_i(0)|_v \leq 1$ and $(\partial P_i / \partial Y)(0, y_i(0)) \neq 0$. Let $|P|_v$ denote the length of the polynomial P with respect to v (see Section 4.2). Then, for any $l \geq 0$, we have the following:*

1. If v is nonarchimedean, then

$$\log^+(|\partial_l(y_1 \cdots y_e)(0)|_v) \leq 2l \max_i \left\{ \log^+ \left(\frac{|P_i|_v}{|(\partial P_i / \partial Y)(0, y_i(0))|_v} \right) \right\}.$$

2. If v is archimedean, then

$$\log^+(|\partial_l(y_1 \cdots y_e)(0)|_v) \leq 2l \max_i \left\{ \log^+ \left(\frac{|P_i|_v}{|(\partial P_i / \partial Y)(0, y_i(0))|_v} \right) \right\}$$

$$+ 7l \max_i \{\log(2 \deg(P_i))\} + \log \binom{l + e - 1}{l}.$$

Proof If $l = 0$, then the assertion is obvious, because $|y_i(0)|_v \le 1$. Further, if $e = 1$, the assertion follows from Lemma 4.31, because $\max\{2l - 1, 0\} \le 2l$.

We assume that $e \ge 2$. By the Leibniz rule, we have

$$\partial_l(y_1 \cdots y_e) = \sum_{l_1 + \cdots + l_e = l} \partial_{l_1}(y_1) \cdots \partial_{l_e}(y_e). \tag{4.24}$$

1. Suppose that v is nonarchimedean. Then it follows from (4.24) that

$$|\partial_l(y_1 \cdots y_e)(0)|_v \le \max_{l_1 + \cdots + l_e = l} \{|\partial_{l_1}(y_1)(0)|_v \cdots |\partial_{l_e}(y_e)(0)|_v\}.$$

Using the case $e = 1$, we have

$$\log^+(|\partial_l(y_1 \cdots y_e)(0)|_v)$$
$$\le \max_{l_1 + \cdots + l_e = l} \{\log^+(|\partial_{l_1}(y_1)(0)|_v) + \cdots + \log^+(|\partial_{l_e}(y_e)(0)|_v)\}$$
$$\le \max_{l_1 + \cdots + l_e = l} \left\{\sum_{i=1}^{e} 2l_i \log^+\left(\frac{|P_i|_v}{|(\partial P_i/\partial Y)(0, y(0))|_v}\right)\right\}$$
$$\le \max_{l_1 + \cdots + l_e = l} \left\{\sum_{i=1}^{e} 2l_i \max_i \left\{\log^+\left(\frac{|P_i|_v}{|(\partial P_i/\partial Y)(0, y_i(0))|_v}\right)\right\}\right\}$$
$$= 2l \max_i \left\{\log^+\left(\frac{|P_i|_v}{|(\partial P_i/\partial Y)(0, y_i(0))|_v}\right)\right\}.$$

2. Suppose that v is archimedean. Then it follows from (4.24) that

$$|\partial_l(y_1 \cdots y_e)(0)|_v \le \binom{l+e-1}{l} \max_{l_1 + \cdots + l_e = l} \{|\partial_{l_1}(y_1)(0)|_v \cdots |\partial_{l_e}(y_e)(0)|_v\}.$$

As in the proof of 1, using the case $e = 1$, we obtain

$$\log^+(|\partial_{l_1}(y_1)(0)|_v \cdots |\partial_{l_e}(y_e)(0)|_v)$$
$$\le 7l \max_i\{\log(2\deg(P_i))\} + 2l \max_i \left\{\log^+\left(\frac{|P_i|_v}{|(\partial P_i/\partial Y)(0, y_i(0))|_v}\right)\right\}.$$

Thus,

$$\log^+(|\partial_l(y_1 \cdots y_e)(0)|_v) \le 2l \max_i \left\{\log^+\left(\frac{|P_i|_v}{|(\partial P_i/\partial Y)(0, y_i(0))|_v}\right)\right\}$$
$$+ 7l \max_i\{\log(2\deg(P_i))\} + \log\binom{l+e-1}{l},$$

as required. $\qquad\square$

5

The Proof of Faltings's Theorem

In this chapter, we give a detailed proof of Faltings's theorem, which asserts that

"any geometrically irreducible algebraic curve of genus at least 2 over a number field has only finitely many rational points."

This is the main theorem of this book and is the culmination of what we have prepared. The proof is elementary but never easy. In Section 5.1, we sketch an outline of the proof. After technical setups required for the proof in Section 5.2, we prove three key results, namely, Theorems 5.4, 5.5, and 5.6, in Sections 5.3, 5.4, and 5.5, respectively. In the last section, we apply Faltings's theorem to Fermat curves.

5.1 Keys for the Proof of Faltings's Theorem

Until Faltings published his proof in 1983, it had not been dreamed that the Mordell conjecture would be solved in the twentieth century. For this achievement, Faltings was awarded the Fields Medal in 1986, and the Mordell conjecture is now called *Faltings's theorem.*

The original proof by Faltings in 1983 used many highly advanced techniques from arithmetic geometry. Subsequently, Vojta (Figure 5.1) [29] gave an alternate, more elementary proof of Faltings's theorem by generalizing techniques from the classical Diophantine approximation to higher dimensions, which was then further generalized by Faltings [9]. However, these new proofs still use a highly nontrivial technique from "Arakelov geometry." Indeed, Vojta used the arithmetic Riemann–Roch theorem for arithmetic threefolds, which had just been established by Gillet and Soulé.

Figure 5.1 Paul A. Vojta.
Source: Archives of the Mathematisches Forschungsinstitut Oberwolfach.

Figure 5.2 Enrico Bombieri.
Source: The Norwegian Academy of Science and Letters

Bombieri (Figure 5.2) [5] succeeded in giving a further elementary proof of Faltings's theorem, replacing this highly nontrivial part with Siegel's lemma (Proposition 4.3). Simply put, Bombieri replaced the arithmetic Riemann–Roch theorem used by Vojta and Faltings with Dirichlet's box principle (the pigeonhole principle).

The purpose of this chapter is to give a detailed proof of Faltings's theorem based on Bombieri's proof. It will be exciting to see how fundamental material from Diophantine geometry in Chapter 4 is effectively used in the proof.

Before we start the proof, we review the classical Diophantine approximation because the proof is achieved by generalizing techniques from this classical Diophantine approximation to higher dimensions.

Let μ be a positive real number and let α be a real algebraic number that is not rational. The question here is whether

$$\left| \alpha - \frac{x}{y} \right| < \frac{1}{y^{\mu}} \tag{5.1}$$

has only finitely many integer solutions (x, y).

For $\mu = 2$, Dirichlet proved that inequality (5.1) has infinitely many integer solutions for any given α. On the other hand, Liouville proved that inequality (5.1) has only finitely many integer solutions if $\mu > [\mathbb{Q}(\alpha) : \mathbb{Q}]$.

Improved bounds for μ were then obtained by Thue, Siegel, Dyson, and others. Ultimately, Roth proved that, if $\mu > 2$, then inequality (5.1) has only finitely many integer solutions for any such given α. For this achievement, Roth was awarded the Fields Medal in 1958. Roth's proof was complicated but elementary, and the key step of the proof was so-called "Roth's lemma" (Theorem 4.20). Indeed, Roth's lemma is also used for the proof of Faltings's theorem (see Section 5.4).

As a prototype of the proof of the Mordell conjecture (in line with the classical Diophantine approximation), we recall a proof of Liouville's theorem here. It is divided roughly into three steps.

A. Find a good polynomial.
B. Consider a certain value using the polynomial in Step A, and give an upper bound for this value.
C. Give a lower bound for this value.

After these steps, we deduce a contradiction by comparing the upper and lower bounds.

We put $d = [\mathbb{Q}(\alpha) : \mathbb{Q}] > 1$.

Step A. For a "good" polynomial, we take the minimal polynomial $P(T) \in \mathbb{Z}[T]$ of α such that the leading coefficient is positive and the greatest common divisor of all the coefficients of $P(T)$ is equal to 1.

Step B. Let $x, y \in \mathbb{Z}$ be integers satisfying $|x/y - \alpha| \leq 1$. As a "certain value," we consider $|P(x/y)|$. Let

$$P(T) = \sum_{i=1}^{d} \frac{P^{(i)}(\alpha)}{i!} (T - \alpha)^i$$

be the Taylor expansion of $P(T)$ at α, and we set

$$M(\alpha) = d \max_{i \geq 1}\{|P^{(i)}(\alpha)/i!|\}.$$

Then, for an upper bound, we obtain

$$|P(x/y)| = \left| \sum_{i=1}^{d} \frac{P^{(i)}(\alpha)}{i!} (x/y - \alpha)^i \right| \leq M(\alpha)|x/y - \alpha|.$$

Step C. Since $P(x/y) \neq 0$, for a lower bound we have $|P(x/y)| \geq 1/y^d$. By combining Step B and Step C, we get

$$\left| \frac{x}{y} - \alpha \right| \geq \frac{1/M(\alpha)}{y^d}$$

for any integers $x, y \in \mathbb{Z}$ with $|x/y - \alpha| \leq 1$. On the other hand, for any sufficiently large y, we have

$$\frac{1/M(\alpha)}{y^d} \geq \frac{1}{y^\mu}.$$

It is now easy to deduce Liouville's theorem.

To generalize this prototype to higher dimensions, we replace the polynomial and the certain value in Steps A–C with a global section of a line bundle and the index of a global section. Step A corresponds to Theorem 5.4, step B to Theorem 5.5, and step C to Theorem 5.6.

For readers who wish to study further, we give a few advanced comments (see the expository books [7] and [30] for more details). Step A is related to Arakelov geometry. Faltings derives general results by using Minkowski's theorem, which are much finer than results obtained by Dirichlet's box principle. Step B is an algebro-geometric part. A general result is Faltings's product theorem [9, Theorem 3.1]. Step C contains involved computations but is not technically difficult.

The purpose of this section is to reduce Faltings's theorem stated below to Theorems 5.4, 5.5, and 5.6. Let K be a number field and $\overline{K}$ be an algebraic closure of K. Let C be a geometrically irreducible, smooth projective curve over K with genus $g \geq 2$.[1]

Theorem 5.1 (Faltings's theorem)   *The set of all K-rational points on C is a finite set.*

After replacing K with a finite field extension of K, we fix a divisor θ on C such that $(2g - 2)\theta$ is linearly equivalent to a canonical divisor ω_C. Note that such θ exists. In fact, we take any $\theta_0 \in C(\overline{K})$, and we set $a = \omega_C - (2g-2)\theta_0 \in \mathrm{Pic}^0(C)(\overline{K})$. Since $\mathrm{Pic}^0(C)(\overline{K})$ is a divisible group (see Corollary 3.25), there is $b \in \mathrm{Pic}^0(C)(\overline{K})$ such that $(2g-2)b = a$. We set $\theta = b + \theta_0$. Then $(2g-2)\theta$ is linearly equivalent to ω_C.

Let J denote the Jacobian variety of C. In the following, we fix the embedding

$$j : C \to J$$

---

[1]   The assumption that C is smooth or projective is not necessary. Indeed, let C be a geometrically irreducible, algebraic curve of (geometric) genus $g \geq 2$ over K. We take a projective curve D over K that contains C as a Zariski open set, and let $\widetilde{D}$ be the normalization of D. Then $\widetilde{D}$ is a geometrically irreducible, smooth projective curve over K with genus $g \geq 2$, and the finiteness of K-rational points on C follows from that on $\widetilde{D}$.

defined by $j(z) = z - \theta$, and identify a point in $C(\overline{K})$ with the corresponding point in $J(\overline{K})$. Let

$$\langle \, , \, \rangle \colon J(\overline{K}) \times J(\overline{K}) \to \mathbb{R},$$

be the Néron–Tate height pairing defined by the theta divisor on J (see Equation (3.13)). Further, for each $x \in J(\overline{K})$, we set

$$|x| = \sqrt{\langle x, x \rangle}.$$

Faltings's theorem is a consequence of Vojta's inequality, which we state below.

Theorem 5.2 (Vojta's inequality)   *Let C be a geometrically irreducible, smooth projective curve over K with genus $g \geq 2$. Let α be a real number with $0 < \alpha < \pi/2$ and $g\cos(\alpha) > \sqrt{g}$. Then there exist positive constants γ, γ' such that, for any $z, z' \in C(K)$ with $|z| \geq \gamma$, $|z'| \geq |z|\gamma'$, we have*

$$\langle z, z' \rangle < \cos(\alpha)|z||z'|.$$

We are going to observe that Theorem 5.2 implies Theorem 5.1.

We set $L = J(K) \otimes_{\mathbb{Z}} \mathbb{R}$. By the Mordell–Weil theorem (Theorem 3.42), L is a finite dimensional $\mathbb{R}$-vector space and, by Proposition 3.33, $\langle \, , \, \rangle$ gives an inner product on L. We consider the composition

$$\tilde{j} \colon C(K) \xrightarrow{\ j\ } J(K) \longrightarrow L = J(K) \otimes \mathbb{R}.$$

We note that each fiber of $\tilde{j}$ is finite. Indeed, by Theorem 3.34, $j \colon C(K) \to J(K)$ is injective. Thus, it suffices to show that each fiber of $J(K) \to J(K) \otimes \mathbb{R}$ is finite. The kernel of $J(K) \to J(K) \otimes \mathbb{R}$ is $J(K)_{\text{tor}}$, i.e., the torsion group consisting of the elements of finite orders in $J(K)$. Since each element in $J(K)_{\text{tor}}$ has height 0, $J(K)_{\text{tor}}$ is a finite group by Northcott's finiteness theorem. Thus, each fiber of $\tilde{j}$ is finite.

We fix a real number α with $0 < \alpha < \pi/2$ and $g\cos(\alpha) > \sqrt{g}$. (For example, if $0 < \alpha < \pi/4$, then it follows from $g \geq 2$ that $\cos(\alpha) > 1/\sqrt{2} \geq 1/\sqrt{g}$. Thus, we can take any number α with $0 < \alpha < \pi/4$.) Let $S = \{x \in L \mid |x| = 1\}$ denote the unit sphere in L. For each $s \in S$, we set

$$\Sigma_s = \{x \in L \mid \langle x, s \rangle \geq \cos(\alpha/2)|x|\}.$$

Then for any $x, x' \in \Sigma_s \setminus \{0\}$, the angle between x and x' is at most α, i.e., we have[2]

---

[2] Indeed, to show (5.3) (see Figure 5.3), we may assume that $x, x' \in \Sigma_s \setminus \{0\}$ belong to S. Then (5.3) is a consequence that $d(s_1, s_2) := \cos^{-1}(\langle s_1, s_2 \rangle) \in [0, \pi]$ satisfies the triangle inequality:

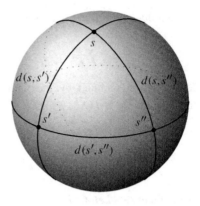

Figure 5.3 Triangle inequality on the sphere.

$$d(s_1, s_2) \leq d(s_1, s_3) + d(s_2, s_3) \qquad (\forall s_1, s_2, s_3 \in S). \qquad (5.2)$$

Indeed, we then have

$$d(x, x') \leq d(x, s) + d(s, x') \leq \alpha/2 + \alpha/2 = \alpha,$$

which gives (5.3). The triangle inequality (5.2) can be shown as follows. We consider the quadratic form

$$Q(x_1, x_2, x_3) = \langle x_1 s_1 + x_2 s_2 + x_3 s_3, x_1 s_1 + x_2 s_2 + x_3 s_3 \rangle$$

on $\mathbb{R}^3$. If we set

$$\langle s_1, s_2 \rangle = \cos(\vartheta_1), \ \langle s_1, s_3 \rangle = \cos(\vartheta_2), \ \langle s_2, s_3 \rangle = \cos(\vartheta_3) \quad (\vartheta_1, \vartheta_2, \vartheta_3 \in [0, \pi]),$$

then the symmetric matrix corresponding to Q is

$$A_Q = \begin{pmatrix} \langle s_1, s_1 \rangle & \langle s_1, s_2 \rangle & \langle s_1, s_3 \rangle \\ \langle s_2, s_1 \rangle & \langle s_2, s_2 \rangle & \langle s_2, s_3 \rangle \\ \langle s_3, s_1 \rangle & \langle s_3, s_2 \rangle & \langle s_3, s_3 \rangle \end{pmatrix} = \begin{pmatrix} 1 & \cos(\vartheta_1) & \cos(\vartheta_2) \\ \cos(\vartheta_1) & 1 & \cos(\vartheta_3) \\ \cos(\vartheta_2) & \cos(\vartheta_3) & 1 \end{pmatrix}.$$

Since $Q(x_1, x_2, x_3) \geq 0$ for any $(x_1, x_2, x_3) \in \mathbb{R}^3$, the eigenvalues of A_Q are all nonnegative. In particular, $\det(A_Q) \geq 0$. On the other hand, since

$$\det(A_Q) = 1 + 2\cos(\vartheta_1)\cos(\vartheta_2)\cos(\vartheta_3) - \cos^2(\vartheta_1) - \cos^2(\vartheta_2) - \cos^2(\vartheta_3),$$

we have

$$(\sin(\vartheta_2)\sin(\vartheta_3))^2 - (\cos(\vartheta_2)\cos(\vartheta_3) - \cos(\vartheta_1))^2$$
$$= (1 - \cos^2(\vartheta_2))(1 - \cos^2(\vartheta_3)) - (\cos(\vartheta_2)\cos(\vartheta_3) - \cos(\vartheta_1))^2 = \det(A_Q) \geq 0.$$

Since $\sin(\vartheta_2)\sin(\vartheta_3) \geq 0$, it follows that

$$\sin(\vartheta_2)\sin(\vartheta_3) \geq |\cos(\vartheta_2)\cos(\vartheta_3) - \cos(\vartheta_1)| \geq \cos(\vartheta_2)\cos(\vartheta_3) - \cos(\vartheta_1),$$

and thus, $\cos(\vartheta_1) \geq \cos(\vartheta_2 + \vartheta_3)$. The required inequality is equivalent to $\vartheta_1 \leq \vartheta_2 + \vartheta_3$, which follows from the above inequality if $\vartheta_2 + \vartheta_3 \leq \pi$ and is trivial if $\vartheta_2 + \vartheta_3 > \pi$.

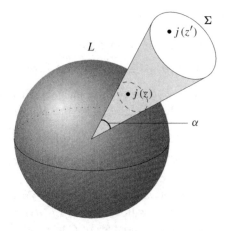

Figure 5.4 Configuration of $j(z)$ and $j(z')$.

$$\langle x, x' \rangle \geq \cos(\alpha)|x||x'|. \tag{5.3}$$

As s is an interior point of Σ_s and S is compact, there are finitely many $s_1, \ldots, s_l \in S$ such that $\Sigma_{s_1} \cup \cdots \cup \Sigma_{s_l} = L$.

To derive Theorem 5.1 from Theorem 5.2, suppose that $C(K)$ is an infinite set. Then $\tilde{j}(C(K))$ is also an infinite set, so there exists s_i such that $\Sigma_{s_i} \cap \tilde{j}(C(K))$ is an infinite set.

We set $\Sigma = \Sigma_{s_i}$. Let γ, γ' be positive constants in Theorem 5.2. Since $\Sigma_{s_i} \cap \tilde{j}(C(K))$ is an infinite set, by Northcott's finiteness theorem, we find an element $z \in C(K)$ with $|z| \geq \gamma$ such that $\tilde{j}(z) \in \Sigma$. Further, by using Northcott's finiteness theorem again, we find an element $z' \in C(K)$ with $|z'| \geq |z|\gamma'$, depending on the previous z, such that $\tilde{j}(z') \in \Sigma$ (Figure 5.4). By (5.3), we then have $\langle z, z' \rangle \geq \cos(\alpha)|z||z'|$, but this inequality contradicts with the conclusion of Theorem 5.2. Thus, Theorem 5.1 is a consequence of Theorem 5.2.

Further, Theorem 5.2 is a consequence of the following three theorems: Theorems 5.4, 5.5, and 5.6. Before stating these theorems, we fix the notation and terminology. We consider the product $C \times C$ and we denote the i-th projection by $p_i : C \times C \to C$ for $i = 1, 2$. Let Δ denote the diagonal divisor on $C \times C$ and we set

$$\Delta' = \Delta - p_1^*(\theta) - p_2^*(\theta).$$

Let d_1, d_2, d be positive integers. We consider the following assumption on d_1, d_2, d:

Assumption 5.3 $gd^2 < d_1 d_2 < g^2 d^2$.

A *Vojta divisor* is a divisor of the form

$$V(d_1, d_2, d) = d_1 p_1^*(\theta) + d_2 p_2^*(\theta) + d\Delta'$$

on $C \times C$ such that d_1, d_2, d are positive integers satisfying Assumption 5.3.
Let s be a nonzero global section of the line bundle $\mathcal{O}_{C \times C}(V(d_1, d_2, d))$:

$$s \in H^0(C \times C, \mathcal{O}_{C \times C}(V(d_1, d_2, d))) \setminus \{0\}.$$

Then we can define the *height* $h(s)$ of s. We will discuss how to define $h(s)$ in Subsection 5.2.2 (see (5.11) and (5.12)).

We can also define the *index* $\mathrm{ind}_{(z, z')}(s; (d_1, d_2))$ of s at a point $(z, z') \in C \times C$ with $z, z' \in C(K)$ with respect to (d_1, d_2), as we now explain. Let ζ and ζ' be local coordinate variables of C at z and z', respectively. (Here, ζ and ζ' may be elements of the completions $\widehat{\mathcal{O}}_{C, z}$ and $\widehat{\mathcal{O}}_{C, z'}$, respectively. See Section 4.3 for details.) Let s_0 be a local frame of $\mathcal{O}_{C \times C}(V(d_1, d_2, d))$ around (z, z'). There is a rational function $f(\zeta, \zeta') \neq 0$ without pole at (z, z') such that $s = f s_0$ locally. Expand $f(\zeta, \zeta')$ into formal power series and write

$$f(\zeta, \zeta') = \sum_{i, j} a_{ij} \zeta^i \zeta'^j.$$

Then, by Proposition 4.11,

$$\min \left\{ \frac{i}{d_1} + \frac{j}{d_2} \,\middle|\, a_{ij} \neq 0 \right\}$$

does not depend on the choices of ζ, ζ', s_0. We denote this quantity by $\mathrm{ind}_{(z, z')}(s; (d_1, d_2))$ and call it the *index* of s at a point $(z, z') \in C \times C$ with respect to (d_1, d_2).

With the above setting, we claim that there are positive constants $c_1, \ldots, c_6$ and a finite subset Z of $C(\overline{K})$ that satisfy Theorems 5.4, 5.5, and 5.6 below.

In the following, $V(d_1, d_2, d)$ denotes a Vojta divisor, i.e., d_1, d_2, d are assumed to satisfy Assumption 5.3. Further, d_1, d_2, d are assumed to satisfy Assumptions 5.9 and 5.10 stated in Section 5.2.

Theorem 5.4 (Existence of a section with small height) *Let C be a geometrically irreducible, smooth projective curve over K with genus $g \geq 2$. Fix a positive number λ_0. Then there is a positive constant c_1 such that, for any sufficiently large d_1, d_2, d satisfying Assumptions 5.3, 5.9, and 5.10 and $d_1 d_2 - gd^2 \geq \lambda_0 d_1 d_2$, there is a nonzero section $s \in H^0(C \times C, \mathcal{O}_{C \times C}(V(d_1, d_2, d)))$ such that*

$$h(s) \leq \frac{c_1}{\lambda_0} \left(1 + \frac{\log(d_1 + d_2 + d)}{d} \right) (d_1 + d_2 + d).$$

(We emphasize that c_1 does not depend on d_1, d_2, d.)

Theorem 5.5 (An upper bound for the indices of sections) *Let C be a geometrically irreducible, smooth projective curve over K with genus $g \geq 2$. Then there exist positive constants c_2, c_3, c_4 and a finite subset $Z \subset C(\overline{K})$ with the following properties: if $\epsilon \in \mathbb{R}$, $d_1, d_2, d \in \mathbb{Z}_{>0}$, $s \in H^0(C \times C, \mathcal{O}_{C \times C}(V(d_1, d_2, d)))$, and $z, z' \in C(K)$ satisfy $0 < \epsilon < 1$, Assumption 5.3, Assumption 5.9, Assumption 5.10, $d_2/d_1 \leq \epsilon^2$, $s \neq 0$, $z, z' \notin Z$, and $\epsilon^2 \min\{d_1|z|^2, d_2|z'|^2\} \geq c_2 h(s) + c_3(d_1 + d_2 + d)$, then we have*

$$\mathrm{ind}_{(z,z')}(s; (d_1, d_2)) \leq c_4 \epsilon.$$

(We emphasize that c_2, c_3, c_4 and $Z \subset C(\overline{K})$ do not depend on ϵ, d_1, d_2, d, s, z, z'.)

Theorem 5.6 (A lower bound for the indices of sections) *Let C be a geometrically irreducible, smooth projective curve over K with genus $g \geq 2$. Then there exist positive constants c_5, c_6 and a finite subset $Z \subset C(\overline{K})$ with the following properties: if $d_1, d_2, d \in \mathbb{Z}_{>0}$, $s \in H^0(C \times C, \mathcal{O}_{C \times C}(V(d_1, d_2, d)))$, and $z, z' \in C(K)$ satisfy Assumption 5.3, Assumption 5.9, Assumption 5.10, $s \neq 0$, and $z, z' \notin Z$, then we have*

$$\mathrm{ind}_{(z,z')}(s; (d_1, d_2))$$
$$\geq \frac{2gd\langle z, z' \rangle - d_1|z|^2 - d_2|z'|^2 - 2gh(s) - c_5(d_1 + d_2 + d)}{c_6 \max\{d_1(1 + |z|^2), d_2(1 + |z'|^2)\}}.$$

(We emphasize that c_5, c_6 and $Z \subset C(\overline{K})$ do not depend on d_1, d_2, d, s, z, z'.)

We remark that the exceptional set Z in Theorem 5.5 and that in Theorem 5.6 can be taken as the same. Indeed, if Z_1 and Z_2 are exceptional sets as in Theorem 5.5 and Theorem 5.6, respectively, then $Z = Z_1 \cup Z_2$ becomes a common exceptional set for Theorem 5.5 and Theorem 5.6.

We are going to show that Theorems 5.4, 5.5, and 5.6 imply Theorem 5.2. As explained in the beginning of this chapter, this will be done by ingeniously comparing upper and lower bounds for the index of a nonzero section s with small height. As the arguments are involved, we divide them into five steps.

Step 1. Given α in Theorem 5.2, we define γ, γ' as follows. First, we fix $\lambda > 0$ with $g \cos(\alpha) - \sqrt{g + \lambda} > 0$ and $\lambda_0 > 0$ with $g^2 \lambda_0 < \lambda$. With respect to λ_0, let c_1 be a constant as in Theorem 5.4. Let $c_2, \ldots, c_6$ and Z be as in Theorem 5.5 and Theorem 5.6. We define a real-valued function $f(t)$ in t by

$$f(t) = \frac{g \cos(\alpha) - \sqrt{g + \lambda}}{c_6 \sqrt{g + \lambda}} - \frac{3}{2} \frac{4gc_1 + c_5 \lambda_0}{c_6 \lambda_0} \frac{1}{t^2}.$$

Since $f(t)$ is positive for any sufficiently large t, we take a sufficiently large positive number c_0 with $f(c_0) > 0$. We take any γ' with

$$\gamma' > \max\left\{1, \frac{c_4}{f(c_0)}\right\}.$$

Further, for this γ', we take any γ such that

$$\gamma > \max\left\{1, c_0, (1+\lambda)\gamma'\sqrt{3\left(\frac{2c_1c_2}{\lambda_0} + c_3\right)}, |x| \,\middle|\, x \in Z\right\}.$$

We are going to observe that the conclusion of Theorem 5.2 holds true for these γ, γ'.

Step 2. Let $z, z' \in C(K)$ be K-rational points satisfying $|z| \geq \gamma$ and $|z'| \geq \gamma'|z|$. To derive a contradiction, we suppose that

$$\langle z, z' \rangle \geq \cos(\alpha)|z||z'|.$$

We first construct a suitable Vojta divisor and a suitable global section associated to the Vojta divisor.

We choose $\lambda_1 > 0$ such that $\sqrt{g + \lambda_1}\frac{|z'|}{|z|}$ is a rational number and $g^2\lambda_0 < \lambda_1 < \lambda$. Further, we choose $\lambda_2 > 0$ such that $\sqrt{g + \lambda_2}\frac{|z|}{|z'|}$ is a rational number, $g^2\lambda_0 < \lambda_2 < \lambda_1$, and $\sqrt{\frac{g+\lambda_1}{g+\lambda_2}} \leq (1+\lambda)^2$. Set

$$a_1 = \sqrt{g + \lambda_1}\frac{|z'|}{|z|}, \qquad a_2 = \sqrt{g + \lambda_2}\frac{|z|}{|z'|}.$$

We fix a positive integer N as in Subsection 5.2.1 (we may take $N = 2g + 1$). Further, we fix a positive integer a such that both aa_1 and aa_2 are integers.

For a sufficiently large integer m that we specify below, we set

$$d_1 = Naa_1m, \quad d_2 = Naa_2m, \quad \text{and} \quad d = Nam.$$

Since $g < \sqrt{g + \lambda_1}\sqrt{g + \lambda_2} = a_1a_2$, we have $gd^2 < d_1d_2$. In addition, as

$$a_1a_2 = \sqrt{g + \lambda_1}\sqrt{g + \lambda_2} < g + \lambda < g^2\cos(\alpha)^2 \leq g^2,$$

we have $d_1d_2 < g^2d^2$. Thus, d_1, d_2, d satisfy Assumption 5.3 and define a Vojta divisor. Further, since

$$\sqrt{g + \lambda_1}\sqrt{g + \lambda_2} - g \geq g + g^2\lambda_0 - g = g^2\lambda_0,$$

we have $d_1d_2 - gd^2 \geq g^2\lambda_0d^2 \geq \lambda_0d_1d_2$.

Moreover, since d_1, d_2, d are multiples of N, they satisfy Assumption 5.9. Note that, if m is sufficiently large, then they also satisfy Assumption 5.10 (this is one of the two conditions that we require for m). Hence d_1, d_2, d can be taken to satisfy all the assumptions of Theorem 5.4.

Note that, if m is taken to be sufficiently large, then

$$\frac{\log(d_1 + d_2 + d)}{d} < 1.$$

(This is the other one of the two conditions that we require for m.) As a consequence, it follows from Theorem 5.4 that there is a nonzero section $s \in H^0(C \times C, \mathcal{O}_{C \times C}(V(d_1, d_2, d)))$ with

$$h(s) \leq \frac{2c_1}{\lambda_0}(d_1 + d_2 + d).$$

Step 3. We give an upper bound of $\mathrm{ind}_{(z,z')}(s; (d_1, d_2))$ by applying Theorem 5.5. We set $\epsilon = 1/\gamma'$. Since γ satisfies $\gamma > \max_{x \in Z}\{|x|\}$ and $z, z' \in C(K)$ satisfy $|z| \geq \gamma$, $|z'| \geq |z|\gamma'$, we have $z, z' \notin Z$. Further,

$$\frac{d_2}{d_1} = \frac{\sqrt{g + \lambda_2}}{\sqrt{g + \lambda_1}}\frac{|z|^2}{|z'|^2} \leq \frac{|z|^2}{|z'|^2} \leq (1/\gamma')^2.$$

To apply Theorem 5.5, let us verify

$$(1/\gamma')^2 \min\{d_1|z|^2, d_2|z'|^2\} \geq c_2 h(s) + c_3(d_1 + d_2 + d). \tag{5.4}$$

First, since $d_2/d_1 \leq (1/\gamma')^2 < 1$, we have $d_1 > d_2$ and the assumption $d_1 d_2 > gd^2$ for the Vojta divisor implies that $d_1 > d$. In particular, $d_1 = \max\{d_1, d_2, d\}$.

Next, if we set

$$u = (1/\gamma')^2 \min\{d_1|z|^2, d_2|z'|^2\} - c_2 h(s) - c_3(d_1 + d_2 + d),$$

then

$$u \geq (1/\gamma')^2 \min\{d\sqrt{g + \lambda_1}|z||z'|, d\sqrt{g + \lambda_2}|z||z'|\}$$
$$\quad - \left(\frac{2c_2 c_1}{\lambda_0} + c_3\right)(d_1 + d_2 + d)$$
$$\geq (1/\gamma')^2 d\sqrt{g + \lambda_2}|z||z'| - 3\left(\frac{2c_2 c_1}{\lambda_0} + c_3\right)d_1$$
$$= \frac{d\sqrt{g + \lambda_2}|z'|}{|z|\gamma'^2}\left(|z|^2 - 3\left(\frac{2c_2 c_1}{\lambda_0} + c_3\right)\gamma'^2\frac{\sqrt{g + \lambda_1}}{\sqrt{g + \lambda_2}}\right)$$
$$\geq \frac{d\sqrt{g + \lambda_2}|z'|}{|z|\gamma'^2}\left(|z|^2 - 3\left(\frac{2c_2 c_1}{\lambda_0} + c_3\right)(\gamma'(1 + \lambda))^2\right)$$
$$\geq \frac{d\sqrt{g + \lambda_2}|z'|}{|z|\gamma'^2}\left(\gamma^2 - 3\left(\frac{2c_2 c_1}{\lambda_0} + c_3\right)(\gamma'(1 + \lambda))^2\right) > 0.$$

Thus, the condition (5.4) is satisfied. Then, by applying Theorem 5.5, we have

$$\text{ind}_{(z,z')}(s;(d_1,d_2)) \leq \frac{c_4}{\gamma'}. \tag{5.5}$$

Step 4. In this step, we give a lower bound for $\text{ind}_{(z,z')}(s;(d_1,d_2))$. By applying Theorem 5.6, we obtain a lower bound

$$\frac{2gd\langle z,z'\rangle - d_1|z|^2 - d_2|z'|^2 - 2g\,h(s) - c_5(d_1 + d_2 + d)}{c_6 \max\{d_1(1 + |z|^2), d_2(1 + |z'|^2)\}}.$$

We are going to simplify this quantity.

First, we estimate the numerator. Since $d_1 > d_2$ and $d_1 > d$, we have

(the numerator)

$$\geq 2gd\langle z,z'\rangle - d_1|z|^2 - d_2|z'|^2 - \frac{4gc_1}{\lambda_0}(d_1 + d_2 + d) - c_5(d_1 + d_2 + d)$$

$$\geq 2gd\langle z,z'\rangle - d_1|z|^2 - d_2|z'|^2 - 3\frac{4gc_1 + \lambda_0 c_5}{\lambda_0}d_1$$

$$\geq 2gd\cos(\alpha)|z||z'| - d\sqrt{g + \lambda_1}|z||z'| - d\sqrt{g + \lambda_2}|z||z'|$$

$$\qquad - 3\frac{4gc_1 + \lambda_0 c_5}{\lambda_0}d\sqrt{g + \lambda_1}\frac{|z'|}{|z|}$$

$$\geq 2d(g\cos(\alpha) - \sqrt{g + \lambda})|z||z'| - 3\frac{4gc_1 + \lambda_0 c_5}{\lambda_0}d\sqrt{g + \lambda}\frac{|z'|}{|z|}$$

$$= 2d\sqrt{g + \lambda}c_6|z||z'|f(|z|).$$

Next, for the denominator, we have

(the denominator)

$$\leq 2c_6 \max\{d_1|z|^2, d_2|z'|^2\}$$

$$= 2c_6 \max\{d\sqrt{g + \lambda_1}|z||z'|, d\sqrt{g + \lambda_2}|z||z'|\} \leq 2d\sqrt{g + \lambda}c_6|z||z'|.$$

Thus,

$$\text{ind}_{(z,z')}(s;(d_1,d_2)) \geq f(|z|). \tag{5.6}$$

Step 5. Since $f(t)$ is monotone increasing for $t > 0$ and $|z| \geq \gamma > c_0$ by our choice of $|z|$ and γ, we have $f(|z|) > f(c_0)$. Combining the upper bound (5.5) and the lower bound (5.6), we obtain

$$f(c_0) < \text{ind}_{(z,z')}(s;(d_1,d_2)) \leq \frac{c_4}{\gamma'}.$$

Thus, $\gamma' f(c_0) < c_4$, which contradicts our choice of γ'.

In conclusion, we have observed that Faltings's theorem is a consequence of Theorems 5.4, 5.5, and 5.6. In the following sections, we prove Theorems 5.4, 5.5, and 5.6.

5.2 Technical Settings for the Proofs of Theorem 5.4, Theorem 5.5, and Theorem 5.6

Let C be a geometrically irreducible, smooth projective curve of genus $g \geq 2$ over a number field K. In the following, we replace K by a finite extension field of K if necessary.

5.2.1 An Embedding C into Projective Space

Let θ be a divisor with $\deg(\theta) = 1$ on C such that $(2g - 2)\theta$ is linearly equivalent to a canonical divisor ω_C on C. As described in Section 5.1, such θ exists if we replace K by a finite extension field of K. Note that θ is ample, so $N\theta$ is very ample for any $N \geq 2g + 1$ (see [11, Chapter IV, Corollary 3.2]). Thus, if $\{\phi_0, \ldots, \phi_n\}$ is a basis $H^0(C, \mathcal{O}_C(N\theta))$, then the morphism $\Phi_{N\theta} : C \to \mathbb{P}^n$ defined by $\Phi_{N\theta}(x) = (\phi_0(x) : \cdots : \phi_n(x))$ is a closed embedding. In the following, we fix such N (for example, we may take $N = 2g + 1$). For $\delta \in \mathbb{Z}$, the pullback of $\mathcal{O}_{\mathbb{P}^N}(\delta)$ by $\Phi_{N\theta} : C \to \mathbb{P}^n$ is denoted by $\mathcal{O}_C(\delta)$, i.e., $\mathcal{O}_C(\delta) = \mathcal{O}_C(\delta N\theta)$. Further, for $\delta, \delta' \in \mathbb{Z}$, the pullback of $\mathcal{O}_{\mathbb{P}^n \times \mathbb{P}^n}(\delta, \delta')$ by $\Phi_{N\theta} \times \Phi_{N\theta} : C \times C \to \mathbb{P}^n \times \mathbb{P}^n$ is denoted by $\mathcal{O}_{C \times C}(\delta, \delta')$. Let $p_i : C \times C \to C$ be the projection to the i-th factor. Then $\mathcal{O}_{C \times C}(\delta, \delta') = p_1^*(\mathcal{O}_C(\delta)) \otimes p_2^*(\mathcal{O}_C(\delta'))$.

If we take homogeneous coordinates $\{X_0, \ldots, X_n\}$ (in other words, a basis $\{\phi_0, \ldots, \phi_n\}$ of $H^0(C, \mathcal{O}_C(N\theta))$) in general, we may assume the following (for details, see [11, Chapter IV, Section 3]):

Assumption 5.7 For any $i \neq j$, $C \cap \{X_i = X_j = 0\} = \emptyset$. Further, for distinct i, j, k, C is birational to the image of C by the projection $\mathbb{P}^n \dashrightarrow \mathbb{P}^2$ given by $(X_0 : \cdots : X_n) \mapsto (X_i : X_j : X_k)$.

For $i \neq j$, we write $\Phi_{N\theta}^{ij}$ for the morphism $C \to \mathbb{P}^1$ induced by the projection $\mathbb{P}^n \dashrightarrow \mathbb{P}^1$, $(X_0 : \cdots : X_n) \mapsto (X_i : X_j)$. Further, for distinct i, j, k, we write $\Phi_{N\theta}^{ijk}$ for the morphism $C \to \mathbb{P}^2$ induced by the projection $\mathbb{P}^n \dashrightarrow \mathbb{P}^2$, $(X_0 : \cdots : X_n) \mapsto (X_i : X_j : X_k)$. Finally, we denote by C_{ijk} the image $\Phi_{N\theta}^{ijk}(C)$. Then

$$(\Phi_{N\theta}^{ij})^*(\mathcal{O}_{\mathbb{P}^1}(1)) = (\Phi_{N\theta}^{ijk})^*(\mathcal{O}_{\mathbb{P}^2}(1)) = \mathcal{O}_C(N\theta).$$

In particular, $\deg(\Phi_{N\theta}^{ij}) = N$ and the degree of C_{ijk} in $\mathbb{P}^2$ is equal to N. Since C does not intersect with $\{X_0 = X_1 = 0\}$, C_{012} does not pass through $(0 : 0 : 1)$, so C_{012} is defined by a homogeneous polynomial of degree N of the form

$$X_2^N + a_{N-1}(X_0, X_1)X_2^{N-1} + \cdots + a_1(X_0, X_1)X_2 + a_0(X_0, X_1),$$

where $a_i(X_0, X_1) \in K[X_0, X_1]$. We choose $c \in O_K \setminus \{0\}$ such that $ca_i(X_0, X_1) \in O_K[X_0, X_1]$ for all i. Multiplying the above homogeneous polynomial by c^N, we have

$$(cX_2)^N + ca_{N-1}(X_0, X_1)(cX_2)^{N-1} + \cdots$$
$$+ c^{N-1}a_1(X_0, X_1)(cX_2) + c^N a_0(X_0, X_1).$$

Replacing X_2 by cX_2 and $c^i a_{N-i}(X_0, X_1)$ by $a_{N-i}(X_0, X_1)$, we may assume the following:

Assumption 5.8 The plane curve C_{012} is defined by a homogeneous polynomial of the form

$$X_2^N + a_{N-1}(X_0, X_1)X_2^{N-1} + \cdots + a_1(X_0, X_1)X_2 + a_0(X_0, X_1)$$

with $a_i(X_0, X_1) \in O_K[X_0, X_1]$ for all $i = 0, \ldots, N - 1$.

5.2.2 Vojta Divisor

Let $p_i : C \times C \to C$ be the projection to the i-th factor, let Δ be the diagonal of $C \times C$, and let $\Delta' = \Delta - p_1^*(\theta) - p_2^*(\theta)$. As explained in Section 5.1, for integers d_1, d_2, d satisfying $gd^2 < d_1d_2 < g^2d^2$, the divisor defined by

$$V(d_1, d_2, d) := d_1 p_1^*(\theta) + d_2 p_2^*(\theta) + d\Delta'$$

is called a *Vojta divisor*. Noting that $p_1^*(\theta) + p_2^*(\theta)$ is an ample divisor, we fix a positive integer t such that $B = tp_1^*(\theta) + tp_2^*(\theta) - \Delta'$ is very ample. Let $\{y_0, \ldots, y_m\}$ be a basis of $H^0(C \times C, \mathcal{O}_{C \times C}(B))$ and $\Phi_{|B|} : C \times C \to \mathbb{P}^m$ be the embedding defined by $\{y_0, \ldots, y_m\}$. Let $(Y_0 : \cdots : Y_m)$ is the homogeneous coordinate of $\mathbb{P}^m$ such that $y_i = Y_i|_{C \times C}$ for all i.

We assume that the integers d_1, d_2, d satisfy the following:

Assumption 5.9 $N \mid (d_1 + td)$ and $N \mid (d_2 + td)$.

Note that if d_1, d_2, d are multiples of N, then Assumption 5.9 holds. We set

$$\delta_1 = \frac{d_1 + td}{N}, \qquad \delta_2 = \frac{d_2 + td}{N}. \tag{5.7}$$

Then δ_1, δ_2 are positive integers, and the Vojta divisor is expressed by the difference of very ample divisors $\delta_1 p_1^*(N\theta) + \delta_2 p_2^*(N\theta)$ and dB, i.e.,

$$V(d_1, d_2, d) = \delta_1 p_1^*(N\theta) + \delta_2 p_2^*(N\theta) - dB. \tag{5.8}$$

We consider the natural linear map

$$H^0(\mathbb{P}^n \times \mathbb{P}^n, \mathcal{O}_{\mathbb{P}^n \times \mathbb{P}^n}(\delta_1, \delta_2)) \longrightarrow H^0(C \times C, \mathcal{O}_{C \times C}(\delta_1, \delta_2)) \tag{5.9}$$

induced by the embedding

$$\Phi_{N\theta} \times \Phi_{N\theta} : C \times C \to \mathbb{P}^n \times \mathbb{P}^n.$$

We further assume that the integers d_1, d_2, d satisfy the following:

Assumption 5.10 The linear map (5.9) is surjective.

Note that Assumption 5.10 holds if d_1, d_2, d are sufficiently large.

Let $(X_0 : \cdots : X_n)$ denote the homogeneous coordinates of $\mathbb{P}^n$ in Section 5.2.1. On $\mathbb{P}^n \times \mathbb{P}^n$, we denote the homogeneous coordinates of the first factor by $(X_0 : \cdots : X_n)$, while we denote the homogeneous coordinates of the second factor by $(X_0' : \cdots : X_n')$ so as not to be confused with the coordinates of the first factor. Then $H^0(\mathbb{P}^n \times \mathbb{P}^n, \mathcal{O}_{\mathbb{P}^n \times \mathbb{P}^n}(\delta_1, \delta_2))$ is isomorphic to the set of bihomogeneous polynomials of bidegree (δ_1, δ_2) in $K[X_0, \ldots, X_n, X_0', \ldots, X_n']$. We denote this space by

$$K[X_0, \ldots, X_n, X_0', \ldots, X_n']_{(\delta_1, \delta_2)}.$$

Further, we set $x_i = X_i|_{C \times C}$ and $x_i' = X_i'|_{C \times C}$. To simplify the notation, we denote

$$K[X_0, \ldots, X_n, X_0', \ldots, X_n'], \ (x_0, \ldots, x_n), \text{ and } (x_0', \ldots, x_n')$$

by $K[X, X']$, x, and x', respectively.

Let U be the linear subspace of $\left(K[X, X']_{(\delta_1, \delta_2)}\right)^{m+1}$ defined by

$$U = \left\{ (F_0, \ldots, F_m) \mid F_0(x, x')/y_0^d = \cdots = F_m(x, x')/y_m^d \right\}. \tag{5.10}$$

Proposition 5.11 *We have a natural surjective linear map*

$$R : U \longrightarrow H^0(C \times C, \mathcal{O}_{C \times C}(V(d_1, d_2, d))).$$

Proof   Let $(F_0, \ldots, F_m) \in U$. For each $i = 0, \ldots, m$, we have

$$\begin{cases} F_i(x, x') \in H^0\left(C \times C, \mathcal{O}_{C \times C}(\delta_1, \delta_2)\right), \\ y_i^d \in H^0(C \times C, \mathcal{O}_{C \times C}(dB)). \end{cases}$$

It follows from (5.8) that $F_i(x, x')/y_i^d$ is a rational section of

$$\mathcal{O}_{C \times C}(V(d_1, d_2, d)).$$

By the definition of U, $F_0(x, x')/y_0^d = \cdots = F_m(x, x')/y_m^d$ on $C \times C$, so $F_i(x, x')/y_i^d$ does not depend on i. Since the pole of $F_i(x, x')/y_i^d$ is contained in $\{y_i = 0\}$, and $y_0, \ldots, y_m$ have no common zeros, we see that $s = F_i(x, x')/y_i^d$ gives rise to a global section of

$$\mathcal{O}_{C \times C}(V(d_1, d_2, d)).$$

Thus, we have a natural linear map

$$R : U \to H^0(C \times C, \mathcal{O}_{C \times C}(V(d_1, d_2, d)))$$

given by $R(F_0, \ldots, F_m) = s$.

To show that R is surjective, let $s \in H^0(C \times C, \mathcal{O}_{C \times C}(V(d_1, d_2, d)))$. Then, for each $i = 0, \ldots, m$, $s y_i^d$ is an element of

$$H^0\left(C \times C, \mathcal{O}_{C \times C}(\delta_1, \delta_2)\right).$$

By Assumption 5.10, there is $F_i(X, X') \in K[X, X']_{(\delta_1, \delta_2)}$ such that $F_i(x, x') = s y_i^d$, i.e.,

$$s = F_0(x, x')/y_0^d = \cdots = F_m(x, x')/y_m^d.$$

Then $(F_0, \ldots, F_m) \in U$ and $s = R(F_0, \ldots, F_m)$. $\square$

Let s be a nonzero global section s of $\mathcal{O}_{C \times C}(V(d_1, d_2, d))$. By Proposition 5.11, there is $(F_0, \ldots, F_m) \in U$ with $R(F_0, \ldots, F_m) = s$. We call $(F_0, \ldots, F_m)$ a *sequence of polynomials expressing* s. Let $h(F_0, \ldots, F_m)$ be the height of the vector given by all the coefficients of $F_0, \ldots, F_m$. Namely, if we write $F_i = \sum a_{i, I, I'} X^I X'^{I'}$ for $i = 0, \ldots, m$, then

$$h(F_0, \ldots, F_m) = \frac{1}{[K : \mathbb{Q}]} \sum_v \log \max_{i, I, I'} \{|a_{i, I, I'}|_v\}. \qquad (5.11)$$

Further, we define the height $h(s)$ of s to be

$$h(s) = \inf_{\substack{(F_0, \ldots, F_m) \in U, \\ R(F_0, \ldots, F_m) = s}} h(F_0, \ldots, F_m). \qquad (5.12)$$

5.2.3 Dimensions of Spaces of Global Sections

We estimate the dimensions of spaces of global sections of some invertible sheaves on $C \times C$.

Lemma 5.12

$$\dim_K H^0(C \times C, \mathcal{O}_{C \times C}(\delta_1, \delta_2)) = (N\delta_1 + 1 - g)(N\delta_2 + 1 - g)$$
$$= N^2 \delta_1 \delta_2 - N(g-1)(\delta_1 + \delta_2) + (g-1)^2.$$

Proof   Since $N \geq 2g+1$ and $\delta_1 \geq 1$, we have $\deg(\mathcal{O}_C(\delta_1)) = N\delta_1 \geq 2g+1$. Then $H^1(C, \mathcal{O}_C(\delta_1)) = 0$, and the Riemann–Roch formula on C gives

$$\dim_K H^0(C, \mathcal{O}_C(\delta_1)) = 1 - g + \deg(\mathcal{O}_C(\delta_1)) = N\delta_1 + 1 - g.$$

We have the same assertion for $\mathcal{O}_C(\delta_2)$. By the projection formula, we have

$$p_{1*}(p_1^*(\mathcal{O}_C(\delta_1)) \otimes p_2^*(\mathcal{O}_C(\delta_2))) = \mathcal{O}_C(\delta_1) \otimes H^0(C, \mathcal{O}_C(\delta_2)).$$

Then

$$H^0(C \times C, \mathcal{O}_{C \times C}(\delta_1, \delta_2)) = H^0(C, \mathcal{O}_C(\delta_1)) \otimes H^0(C, \mathcal{O}_C(\delta_2)),$$

and the assertion follows.   □

Lemma 5.13   *Let $V = V(d_1, d_2, d)$ be a Vojta divisor. Then we have the following:*

1. *If $d_1 + d_2 > 4g - 4$, then $H^2(C \times C, \mathcal{O}_{C \times C}(V)) = 0$.*
2. *If $d_1 + d_2 > 4g - 4$, then*

$$\dim_K H^0(C \times C, \mathcal{O}_{C \times C}(V)) \geq (d_1 d_2 - g d^2) - (g-1)(d_1 + d_2) + \chi(\mathcal{O}_{C \times C}).$$

Proof   1. Let $\omega_{C \times C}$ be a canonical divisor of $C \times C$, i.e., $\omega_{C \times C} = p_1^*(\omega_C) + p_2^*(\omega_C)$. We set $D := p_1^*(\theta) + p_2^*(\theta)$. Recall that $V = (d_1 - d)p_1^*(\theta) + (d_2 - d)p_2^*(\theta) + d\Delta$. Using

$$p_1^*(\theta) \cdot p_1^*(\theta) = p_2^*(\theta) \cdot p_2^*(\theta) = 0, \ p_1^*(\theta) \cdot p_2^*(\theta) = 1, \ \Delta \cdot p_1^*(\theta) = \Delta \cdot p_2^*(\theta) = 1,$$

we obtain $(\omega_{C \times C} - V) \cdot D = (4g - 4) - d_1 - d_2 < 0$. Since D is ample, we have $H^2(C \times C, \mathcal{O}_{C \times C}(\omega_{C \times C} - V)) = 0$. By Serre's duality theorem, we obtain $H^0(C \times C, \mathcal{O}_{C \times C}(V)) = 0$.

2. It follows from $\Delta \cdot \Delta = 2 - 2g$ that $V \cdot V = 2d_1 d_2 - 2g d^2$. Further, it follows from $\omega_{C \times C} \cdot \Delta = 4g - 4$ that $V \cdot \omega_{C \times C} = 2(g-1)(d_1 + d_2)$. Then,

by 1 and the Riemann–Roch formula on $C \times C$, we have

$$\dim_K H^0(C \times C, \mathcal{O}_{C \times C}(V)) \geq \chi(\mathcal{O}_{C \times C}(V))$$
$$= \frac{(V \cdot (V - \omega_{C \times C}))}{2} + \chi(\mathcal{O}_{C \times C})$$
$$= (d_1 d_2 - g d^2) - (g - 1)(d_1 + d_2) + \chi(\mathcal{O}_{C \times C}),$$

as desired. □

5.3 Existence of Small Section (the Proof of Theorem 5.4)

As explained at the beginning of this chapter, the alternate proof of Faltings's theorem by Vojta [29] uses the arithmetic Riemann–Roch theorem in Arakelov geometry to ensure the existence of a small section. Then Bombieri [5] found a method to circumvent the arithmetic Riemann–Roch formula, replacing it with Siegel's lemma. The proof of Theorem 5.4 in this section is based on Bombieri's method.

We first give a rough sketch of the proof of Theorem 5.4. Our goal is to find a nonzero section s of $H^0(C \times C, \mathcal{O}_{C \times C}(V(d_1, d_2, d)))$ with "small" height.

We use the natural surjective homomorphism

$$R: U \longrightarrow H^0(C \times C, \mathcal{O}_{C \times C}(V(d_1, d_2, d)))$$

in Proposition 5.11, where U is the subspace of $\left(K[X, X']_{(\delta_1, \delta_2)} \right)^{m+1}$ defined by

$$U = \left\{ (F_0, \ldots, F_m) \mid F_0(x, x')/y_0^d = \cdots = F_m(x, x')/y_m^d \right\}.$$

To give an element of U amounts to solving the system of linear equations $F_i(x, x')y_j^d = F_j(x, x')y_i^d$ for $0 \leq i, j \leq m$. We reduce the above system of linear equations to a system of linear equations with coefficients in O_K. (To be precise, to ease computation, we consider a linear subspace of U which is defined by linear equations with coefficients in O_K.) Then we apply Siegel's lemma, and we obtain a nonzero element $(F_0, \ldots, F_m)$ of U with "small" height such that $R(F_0, \ldots, F_m) \neq 0$. Then $s = R(F_0, \ldots, F_m)$ gives a desired section of $H^0(C \times C, \mathcal{O}_{C \times C}(V(d_1, d_2, d)))$ in Theorem 5.4.

We now carry out the details. We divide the proof into six steps.

Step 1. First, we simplify the y_i's. Recall that we fix a positive integer t in Subsection 5.2.2. Let t' be a sufficient large integer such that $t + 1 + t'$ is divisible by N. We set $t + 1 + t' = Nr$. Since r is sufficiently large,

$$H^0(\mathbb{P}^n \times \mathbb{P}^n, \mathcal{O}_{\mathbb{P}^n \times \mathbb{P}^n}(r,r)) \longrightarrow H^0(C \times C, \mathcal{O}_{C \times C}(r,r))$$
$$= H^0(C \times C, \mathcal{O}_{C \times C}(Nr(p_1^*(\theta) + p_2^*(\theta)))) \tag{5.13}$$

is surjective.

Claim 1 There exist a positive integer r' and nonzero bihomogeneous polynomials $P_0, \ldots, P_m, Q \in O_K[X_0, X_1, X_2, X_0', X_1', X_2']$ of bidegree (r', r') with the following property:

For any

$$F_0, \ldots, F_m \in K[X, X']_{(\delta_1, \delta_2)},$$

we have

$$F_i(x, x')y_j^d = F_j(x, x')y_i^d \quad \text{if and only if}$$
$$F_i(x, x')(P_j(x_{012}, x'_{012})Q(x_{012}, x'_{012}))^d$$
$$= F_j(x, x')(P_i(x_{012}, x'_{012})Q(x_{012}, x'_{012}))^d$$

for all $0 \leq i \leq m$, $0 \leq j \leq m$, where $x_{012} = (x_0, x_1, x_2)$ and $x'_{012} = (x_0', x_1', x_2')$.

Proof Let ϕ be a canonical section of $H^0(C \times C, \mathcal{O}_{C \times C}(\Delta))$ (i.e., a section with $\mathrm{div}(\phi) = \Delta$). Then, since

$$B = tp_1^*(\theta) + tp_2^*(\theta) - \Delta' = (t+1)p_1^*(\theta) + (t+1)p_2^*(\theta) - \Delta$$

and $y_i \in H^0(C \times C, \mathcal{O}_{C \times C}(B))$, ϕy_i is an element of

$$H^0(C \times C, \mathcal{O}_{C \times C}((t+1)(p_1^*(\theta) + p_2^*(\theta)))).$$

Let ϕ' be a nonzero element of $H^0(C, \mathcal{O}_C(t'\theta))$. Then $\phi y_i p_1^*(\phi')p_2^*(\phi') \in H^0(C \times C, \mathcal{O}_{C \times C}(r,r))$. Since the linear map (5.13) is surjective, there is a bihomogeneous polynomial $L_i(X, X')$ of bidegree (r, r) such that $L_i(x, x') = \phi y_i p_1^*(\phi')p_2^*(\phi')$. Then

$$F_i(x, x')y_j^d = F_j(x, x')y_i^d \iff F_i(x, x')L_j(x, x')^d = F_j(x, x')L_i(x, x')^d.$$

Let C' be the image of C by the projection $(X_0 : \cdots : X_n) \mapsto (X_0 : X_1 : X_2)$ (C' was denoted by C_{012} in Subsection 5.2.1). Then C is birational to C' by Assumption 5.7. On the other hand,

$$L_i(1, x_1/x_0, \ldots, x_n/x_0, 1, x_1'/x_0', \ldots, x_n'/x_0')$$
$$\in K(C \times C) = K(C' \times C') = K(x_1/x_0, x_2/x_0, x_1'/x_0', x_2'/x_0'),$$

i.e., there are $p_i, q_i \in K[X_1/X_0, X_2/X_0, X_1'/X_0', X_2'/X_0']$ such that

$$L_i(1, x_1/x_0, \ldots, x_n/x_0, 1, x_1'/x_0', \ldots, x_n'/x_0') = \frac{p_i(x_1/x_0, x_2/x_0, x_1'/x_0', x_2'/x_0')}{q_i(x_1/x_0, x_2/x_0, x_1'/x_0', x_2'/x_0')}.$$

Replacing q_i by $q_0 \cdots q_m$, and p_i by $p_i q_0 \cdots q_{i-1} \cdot q_{i+1} \cdots q_m$ for each $i = 0, \ldots, m$, we may assume that $q_0 = \cdots = q_m$. We denote q_0 by q. Then there exist bihomogeneous polynomials $P_0, \ldots, P_m, Q \in K[X_0, X_1, X_2, X_0', X_1', X_2']$ such that

$$L_i(1, x_1/x_0, \ldots, x_n/x_0, 1, x_1'/x_0', \ldots, x_n'/x_0')$$
$$= \frac{P_i(1, x_1/x_0, x_2/x_0, 1, x_1'/x_0', x_2'/x_0')}{Q(1, x_1/x_0, x_2/x_0, 1, x_1'/x_0', x_2'/x_0')}.$$

We note that if we set $\tilde{P}_i = X_0^{e_i} X_0'^{e_i'} P_i$ and $\tilde{Q} = X_0^f X_0'^{f'} Q$ ($e_i, e_i', f, f' \in \mathbb{Z}_{\geq 0}$), then

$$\begin{cases} \tilde{P}_i(1, x_1/x_0, x_2/x_0, 1, x_1'/x_0', x_2'/x_0') = P_i(1, x_1/x_0, x_2/x_0, 1, x_1'/x_0', x_2'/x_0'), \\ \tilde{Q}(1, x_1/x_0, x_2/x_0, 1, x_1'/x_0', x_2'/x_0') = Q(1, x_1/x_0, x_2/x_0, 1, x_1'/x_0', x_2'/x_0'). \end{cases}$$

Thus, we can adjust the bidegrees of P_i, Q by choosing suitable $e_i, e_i', f, f' \in \mathbb{Z}_{\geq 0}$, so we may assume that there is a positive integer r' such that $P_0, \ldots, P_m, Q \in K[X_0, X_1, X_2, X_0', X_1', X_2']$ are all of bidegree (r', r'). Further, replacing $P_0, \ldots, P_m, Q$ by $a P_0, \ldots, a P_m, a Q$ for some $a \in O_K \setminus \{0\}$, we may assume that

$$P_0, \ldots, P_m, Q \in O_K[X_0, X_1, X_2, X_0', X_1', X_2'].$$

This completes the proof of the claim. $\square$

Step 2. We define a homomorphism

$$\alpha \colon H^0(C \times C, \mathcal{O}_{C \times C}(\delta_1, \delta_2))^{m+1}$$
$$\longrightarrow H^0(C \times C, \mathcal{O}_{C \times C}(\delta_1 + 2r'd, \delta_2 + 2r'd))^{m(m+1)/2}$$

by

$$(f_0, \ldots, f_m) \mapsto \left(f_i(Q(x_{012}, x'_{012}) P_j(x_{012}, x'_{012}))^d \right.$$
$$\left. - f_j(Q(x_{012}, x'_{012}) P_i(x_{012}, x'_{012}))^d \right)_{i < j}.$$

By Proposition 5.11 and Claim 1, $H^0(C \times C, \mathcal{O}_{C \times C}(V(d_1, d_2, d))) = \mathrm{Ker}(\alpha)$.

As it is difficult to find a concrete basis of $H^0(C \times C, \mathcal{O}_{C \times C}(\delta_1, \delta_2))$, we consider the subspace

$$K[x_0, x_1, x_2, x_0', x_1', x_2']_{(\delta_1, \delta_2)}$$
$$:= \{\beta(F) \mid F \in K[X_0, X_1, X_2, X_0', X_1', X_2']_{(\delta_1, \delta_2)}\},$$

of $H^0(C \times C, \mathcal{O}_{C \times C}(\delta_1, \delta_2))$, where $K[X_0, X_1, X_2, X_0', X_1', X_2']_{(\delta_1, \delta_2)}$ is the space consisting of bihomogenous polynomials of bidegree (δ_1, δ_2) and

$$\beta : K[X_0, X_1, X_2, X_0', X_1', X_2'] \to K[x_0, x_1, x_2, x_0', x_1', x_2']$$

is the canonical homomorphism.

We write

$$\alpha' : K[x_0, x_1, x_2, x_0', x_1', x_2']_{(\delta_1, \delta_2)}^{m+1}$$
$$\longrightarrow H^0(C \times C, \mathcal{O}_{C \times C}(\delta_1 + 2r'd, \delta_2 + 2r'd))^{m(m+1)/2}$$

for the restriction of α to $K[x_0, x_1, x_2, x_0', x_1', x_2']_{(\delta_1, \delta_2)}^{m+1}$. Note that, if $f_0 = \beta(F_0), \ldots, f_m = \beta(F_m)$ with

$$F_0, \ldots, F_m \in K[X_0, X_1, X_2, X_0', X_1', X_2']_{(\delta_1, \delta_2)},$$

then

$$\alpha'(f_0, \ldots, f_m) = \Big(F_i(x_{012}, x'_{012})(Q(x_{012}, x'_{012})P_j(x_{012}, x'_{012}))^d$$
$$- F_j(x_{012}, x'_{012})(Q(x_{012}, x'_{012})P_i(x_{012}, x'_{012}))^d \Big)_{i<j}.$$

Thus, the image of α' is contained in

$$K[x_0, x_1, x_2, x_0', x_1', x_2']_{(\delta_1 + 2r'd, \delta_2 + 2r'd)}^{m(m+1)/2},$$

so we write

$$\alpha' : K[x_0, x_1, x_2, x_0', x_1', x_2']_{(\delta_1, \delta_2)}^{m+1}$$
$$\longrightarrow K[x_0, x_1, x_2, x_0', x_1', x_2']_{(\delta_1 + 2r'd, \delta_2 + 2r'd)}^{m(m+1)/2}. \quad (5.14)$$

Note that $\mathrm{Ker}(\alpha') \subseteq \mathrm{Ker}(\alpha) = H^0(C \times C, \mathcal{O}_{C \times C}(V(d_1, d_2, d)))$. In order to find a suitable section of $H^0(C \times C, \mathcal{O}_{C \times C}(V(d_1, d_2, d)))$ for Theorem 5.4, we consider elements of $\mathrm{Ker}(\alpha')$ in the following:

Step 3. Let us compute the dimension of $K[x_0, x_1, x_2, x_0', x_1', x_2']_{(\delta_1, \delta_2)}$ by finding a basis.

Claim 2   Let $H(\delta_1, \delta_2)$ be the subspace of $K[X_0, X_1, X_2, X_0', X_1', X_2']$ consisting of bihomogeneous polynomials of bidegree (δ_1, δ_2) of the form

$$\sum_{\substack{0 \leq c < N \\ 0 \leq c' < N}} q_{cc'}(X_0, X_1, X_0', X_1') X_2^c X_2'^{c'}.$$

Then the restriction of

$$\beta : K[X_0, X_1, X_2, X_0', X_1', X_2'] \to K[x_0, x_1, x_2, x_0', x_1', x_2']$$

to $H(\delta_1, \delta_2)$ induces the isomorphism

$$H(\delta_1, \delta_2) \to K[x_0, x_1, x_2, x_0', x_1', x_2']_{(\delta_1, \delta_2)}.$$

Proof   By Assumption 5.8, C' is defined by

$$X_2^N + a_{N-1}(X_0, X_1) X_2^{N-1} + \cdots + a_1(X_0, X_1) X_2 + a_0(X_0, X_1) = 0,$$

from which the assertion follows. $\square$

By Claim 2, if $\delta_1 \geq N, \delta_2 \geq N$, then

$$\left\{ x_0^a x_1^b x_2^c x_0'^{a'} x_1'^{b'} x_2'^{c'} \right\}_{\substack{0 \leq c < N \\ a+b=\delta_1-c \\ 0 \leq c' < N \\ a'+b'=\delta_2-c'}}$$

forms a basis of $K[x_0, x_1, x_2, x_0', x_1', x_2']_{(\delta_1, \delta_2)}$. Note that

$$\# \left\{ (a,b,c) \in \mathbb{Z}_{\geq 0}^3 \mid 0 \leq c < N, \ a+b = \delta_1 - c \right\}$$

$$= \sum_{c=0}^{N-1} (\delta_1 - c + 1) = N\delta_1 - \frac{N(N-3)}{2}$$

and a similar estimate holds for (a', b', c'). We obtain

$$\dim_K K[x_0, x_1, x_2, x_0', x_1', x_2']_{(\delta_1, \delta_2)}$$

$$= \left(N\delta_1 - \frac{N(N-3)}{2} \right) \left(N\delta_2 - \frac{N(N-3)}{2} \right)$$

$$= N^2 \delta_1 \delta_2 - \frac{N^2(N-3)}{2} (\delta_1 + \delta_2) + \left(\frac{N(N-3)}{2} \right)^2. \quad (5.15)$$

Step 4. Next let us see that $\dim_K \mathrm{Ker}(\alpha')$ is relatively large. As vector spaces over K, we have

$$\mathrm{Ker}(\alpha') = \mathrm{Ker}(\alpha) \cap K[x_0, x_1, x_2, x_0', x_1', x_2']_{(\delta_1, \delta_2)}^{m+1},$$

so we have

$$\dim_K \text{Ker}(\alpha') = \dim_K \text{Ker}(\alpha) + \dim_K K[x_0,x_1,x_2,x'_0,x'_1,x'_2]^{m+1}_{(\delta_1,\delta_2)}$$
$$- \dim_K \left(\text{Ker}(\alpha) + K[x_0,x_1,x_2,x'_0,x'_1,x'_2]^{m+1}_{(\delta_1,\delta_2)} \right).$$

Since

$$\text{Ker}(\alpha) + K[x_0,x_1,x_2,x'_0,x'_1,x'_2]^{m+1}_{(\delta_1,\delta_2)} \subseteq H^0(C \times C, \mathcal{O}_{C \times C'}(\delta_1,\delta_2))^{m+1},$$

we have

$$\dim_K \text{Ker}(\alpha') \geq \dim_K \text{Ker}(\alpha) - \left(\dim_K H^0(C \times C, \mathcal{O}_{C \times C'}(\delta_1,\delta_2))^{m+1} \right.$$
$$\left. - \dim_K K[x_0,x_1,x_2,x'_0,x'_1,x'_2]^{m+1}_{(\delta_1,\delta_2)} \right).$$

By Lemma 5.13, we can take a positive constant B_1 independent of d,d_1,d_2 (for example, we can take $B_1 = 2(g-1) + \chi(\mathcal{O}_{C \times C})$) such that, for any sufficiently large d_1, d_2, d satisfying the assumptions in Theorem 5.4,

$$\dim_K \text{Ker}(\alpha) = \dim_K H^0(C \times C, \mathcal{O}_{C \times C}(V(d_1,d_2,d)))$$
$$\geq (d_1 d_2 - g d^2) - (g-1)(d_1 + d_2) + \chi(\mathcal{O}_{C \times C})$$
$$\geq \lambda_0 d_1 d_2 - B_1(d_1 + d_2).$$

Further, by Lemma 5.12 and (5.15), we can take a positive constant B_2 independent of d,d_1,d_2 such that

$$\dim_K H^0(C \times C, \mathcal{O}_{C \times C}(\delta_1,\delta_2))^{m+1} - \dim_K K[x_0,x_1,x_2,x'_0,x'_1,x'_2]^{m+1}_{(\delta_1,\delta_2)}$$
$$\leq (m+1)\left(\frac{N(N-1)}{2} - (g-1) \right)(d_1 + d_2 + 2td)$$
$$+ (m+1)\left((g-1)^2 - \left(\frac{N(N-3)}{2} \right)^2 \right)$$
$$\leq B_2(d_1 + d_2 + d).$$

Since $d_1 d_2 > g d^2$, we have

$$\dim_K \text{Ker}(\alpha') \geq \lambda_0 d_1 d_2 - B_1(d_1 + d_2) - B_2(d_1 + d_2 + d)$$
$$> \frac{\lambda_0 d_1 d_2}{2} + \left(\frac{\lambda_0 g d^2}{4} - B_2 d \right) + \left(\frac{\lambda_0 d_1 d_2}{4} - (B_1 + B_2)(d_1 + d_2) \right)$$
$$= \frac{\lambda_0 d_1 d_2}{2} + \frac{\lambda_0 g d}{4}\left(d - \frac{4 B_2}{\lambda_0 g} \right)$$
$$+ \frac{\lambda_0}{4}\left\{ \left(d_1 - \frac{4(B_1 + B_2)}{\lambda_0} \right)\left(d_2 - \frac{4(B_1 + B_2)}{\lambda_0} \right) - \left(\frac{4(B_1 + B_2)}{\lambda_0} \right)^2 \right\}.$$

Thus, for any sufficiently large d_1, d_2, d satisfying the assumptions in Theorem 5.4, we obtain

$$\dim_K \operatorname{Ker}(\alpha') \geq \frac{\lambda_0 d_1 d_2}{2}. \tag{5.16}$$

Step 5. Let A' be the matrix representation of α' with respect to the bases

$$\left\{ x_0^a x_1^b x_2^c x_0'^{a'} x_1'^{b'} x_2'^{c'} \right\}_{\substack{0 \leq c < N \\ a+b=\delta_1-c \\ 0 \leq c' < N \\ a'+b'=\delta_2-c'}} \quad \text{and} \quad \left\{ x_0^a x_1^b x_2^c x_0'^{a'} x_1'^{b'} x_2'^{c'} \right\}_{\substack{0 \leq c < N \\ a+b=\delta_1+2r'd-c \\ 0 \leq c' < N \\ a'+b'=\delta_2+2r'd-c'}}$$

of $K[x_0, x_1, x_2, x_0', x_1', x_2']_{(\delta_1, \delta_2)}$ and $K[x_0, x_1, x_2, x_0', x_1', x_2']_{(\delta_1+2r'd, \delta_2+2r'd)}$ (see (5.14) and Step 3). We are going to evaluate the entries of A'. We set

$$F_i(X_0, X_1, X_2, X_0', X_1', X_2') = \sum_{\substack{0 \leq c < N \\ a+b=\delta_1-c \\ 0 \leq c' < N \\ a'+b'=\delta_2-c'}} T_{i,abca'b'c'} X_0^a X_1^b X_2^c X_0'^{a'} X_1'^{b'} X_2'^{c'},$$

where $T_{i,abca'b'c'}$'s are unknown variables to which we later apply Siegel's lemma to find a nonzero small section. Then

$$F_i(\boldsymbol{x}_{012}, \boldsymbol{x}'_{012})(Q(\boldsymbol{x}_{012}, \boldsymbol{x}'_{012}) P_j(\boldsymbol{x}_{012}, \boldsymbol{x}'_{012}))^d$$
$$= \sum_{\substack{0 \leq c < N \\ a+b=\delta_1-c \\ 0 \leq c' < N \\ a'+b'=\delta_2-c'}} T_{i,abca'b'c'} x_0^a x_1^b x_0'^{a'} x_1'^{b'} \left(x_2^c x_2'^{c'} (Q(\boldsymbol{x}_{012}, \boldsymbol{x}'_{012}) P_j(\boldsymbol{x}_{012}, \boldsymbol{x}'_{012}))^d \right).$$

Since $a_0, \ldots, a_{N-1} \in O_K[X_0, X_1]$, by Proposition 4.29, there are polynomials

$$h_{ll'}^{jcc'}(S, S') \in O_K[S, S']$$

of bidegree at most $(c + 2dr' - l, c' + 2dr' - l')$ such that if we set $\boldsymbol{x}_{012}/x_0 = (1, x_1/x_0, x_2/x_0)$ and $\boldsymbol{x}'_{012}/x_0' = (1, x_1'/x_0', x_2'/x_0')$, then

$$(x_2/x_0)^c (x_2'/x_0')^{c'} (Q(\boldsymbol{x}_{012}/x_0, \boldsymbol{x}'_{012}/x_0') P_j(\boldsymbol{x}_{012}/x_0, \boldsymbol{x}'_{012}/x_0'))^d$$
$$= \sum_{\substack{0 \leq l < N \\ 0 \leq l' < N}} h_{ll'}^{jcc'}(x_1/x_0, x_1'/x_0')(x_2/x_0)^l (x_2'/x_0')^{l'}. \tag{5.17}$$

Further, for each place v, there is a constant c_v depending only on $a_0, \ldots, a_{N-1}$ such that

$$|h_{ll'}^{jcc'}(S, S')|_v \leq |T^c T'^{c'} (Q(1, S, T, 1, S', T') P_j(1, S, T, 1, S', T'))^d|_v$$
$$\times \exp(c_v(c + c' + 4dr')). \tag{5.18}$$

Here, we set

$$H_{ll'}^{jcc'}(X_0, X_1, X_0', X_1') = X_0^{c+2dr'-l} X_0'^{c'+2dr'-l'} h_{ll'}^{jcc'}(X_1/X_0, X_1'/X_0').$$

Then $H_{ll'}^{jcc'}(X_0, X_1, X_0', X_1') \in O_K[X_0, X_1, X_0', X_1']$ and it is either zero or a nonzero bihomogeneous polynomial of bidegree $(c + 2dr' - l, c' + 2dr' - l')$. Further, multiplying (5.17) by $x_0^{c+2dr'} x_0'^{c'+2dr'}$, we have

$$x_2^c x_2'^{c'} (Q(\boldsymbol{x}_{012}, \boldsymbol{x}'_{012}) P_j(\boldsymbol{x}_{012}, \boldsymbol{x}'_{012}))^d$$
$$= \sum_{\substack{0 \le l < N \\ 0 \le l' < N}} H_{ll'}^{jcc'}(x_0, x_1, x_0', x_1') x_2^l x_2'^{l'}. \quad (5.19)$$

Thus, the difference of the coefficients of $H_{ll'}^{jcc'}$ and $H_{ll'}^{icc'}$ are entries of A'. Further, by (5.18),

$$|H_{ll'}^{jcc'}(X_0, X_1, X_0', X_1')|_v$$
$$= |H_{ll'}^{jcc'}(1, X_1, 1, X_1')|_v = |h_{ll'}^{jcc'}(S, S')|_v$$
$$\le |T^c T'^{c'} (Q(1, S, T, 1, S', T') P_j(1, S, T, 1, S', T'))^d|_v \exp(c_v(c + c' + 4dr'))$$
$$= |(Q(1, S, T, 1, S', T') P_j(1, S, T, 1, S', T'))^d|_v \exp(c_v(c + c' + 4dr'))$$
$$\le |(Q(1, S, T, 1, S', T') P_j(1, S, T, 1, S', T'))^d|_v \exp(c_v(4dr' + 2(N - 1))).$$

Now we assume that v is archimedean. Since

$$Q(1, S, T, 1, S', T') P_j(1, S, T, 1, S', T')$$

is a polynomial of bidegree at most $(2r', 2r')$ with four variables, Corollary 4.6 tells us that

$$|(Q(1, S, T, 1, S', T') P_j(1, S, T, 1, S', T'))^d|_v$$
$$\le (1 + 2r')^{4(d-1)} |Q(1, S, T, 1, S', T') P_j(1, S, T, 1, S', T')|_v^d.$$

Thus, there is a constant A_1 independent of d_1, d_2, d such that, for all i, j, c, c', l, l' and all archimedean place v, we have

$$|H_{ll'}^{jcc'}|_v \le \exp(A_1 d).$$

In conclusion, for any entry e of A', we obtain that $e \in O_K$ and

$$\|e\|_K \le 2 \exp(A_1 d),$$

where $\|e\|_K = \max_{v \in M_K^\infty} \{|e|_v\}$ as defined in (4.1).

Step 6. By Siegel's lemma (Proposition 4.3), there are constants A_2 and A_3 depending only on K and the choice of a free basis O_K over $\mathbb{Z}$ such that we can find nontrivial solutions $t_{i,abca'b'c'} \in O_K$ for unknown variables $T_{i,abca'b'c'}$ with

$$\max_{i,a,b,c,a',b',c'} \log \|t_{i,abca'b'c'}\|_K \leq \frac{\mathrm{rk}\,\alpha'}{\dim \mathrm{Ker}(\alpha')} \times$$
$$\left(A_1 d + \log \left(\dim_K K[x_0,x_1,x_2,x_0',x_1',x_2']_{(\delta_1,\delta_2)}^{m+1} \right) + A_2 \right) + A_3.$$

By using (5.15) and $\delta_1 = \frac{d_1+dt}{N}, \delta_2 = \frac{d_2+dt}{N}$, if d_1, d_2, d are sufficiently large, then

$$\dim_K K[x_0,x_1,x_2,x_0',x_1',x_2']_{(\delta_1,\delta_2)}^{m+1} \leq (m+1)(d_1+dt)(d_2+dt),$$

so there is a constant A_4 independent from d_1, d_2, d such that

$$A_1 d + \log \left(\dim_K K[x_0,x_1,x_2,x_0',x_1',x_2']_{(\delta_1,\delta_2)}^{m+1} \right) + A_2 \leq$$
$$A_4 d \left(1 + \frac{\log(d_1+d_2+d)}{d} \right).$$

Let us evaluate $\dfrac{\mathrm{rk}\,\alpha'}{\dim \mathrm{Ker}(\alpha')} d$. Since $gd^2 < d_1 d_2$ by Assumption 5.3, we have $d^2 < g^2 d_1 d_2$, so by (5.16),

$$\frac{\mathrm{rk}\,\alpha'}{\dim \mathrm{Ker}(\alpha')} d \leq \frac{\dim_K K[x_0,x_1,x_2,x_0',x_1',x_2']_{(\delta_1,\delta_2)}^{m+1}}{\dim \mathrm{Ker}(\alpha')} d$$
$$\leq \frac{(m+1)(d_1+td)(d_2+td)}{\lambda_0 d_1 d_2/2} d$$
$$= \frac{2(m+1)}{\lambda_0} \left\{ d + t(d_1+d_2)\frac{d^2}{d_1 d_2} + t^2 d \frac{d^2}{d_1 d_2} \right\}$$
$$\leq \frac{2(m+1)}{\lambda_0} \left\{ d + t(d_1+d_2)g^2 + t^2 dg^2 \right\}$$
$$\leq \frac{2(m+1)}{\lambda_0} (1 + tg^2 + t^2 g^2)(d_1 + d_2 + d).$$

We set $A_5 = 2(m+1)(1 + tg^2 + t^2 g^2)$. Then

$$\frac{\mathrm{rk}\,\alpha'}{\dim \mathrm{Ker}(\alpha')} d \leq \frac{A_5}{\lambda_0}(d_1 + d_2 + d),$$

so

$$\max_{i,a,b,c,a',b',c'} \log \|t_{i,abca'b'c'}\|_K \leq \frac{\mathrm{rk}\,\alpha'}{\dim \mathrm{Ker}(\alpha')} d A_4 \left(1 + \frac{\log(d_1+d_2+d)}{d} \right) + A_3$$
$$\leq \frac{A_4 A_5}{\lambda_0}(d_1 + d_2 + d) \left(1 + \frac{\log(d_1+d_2+d)}{d} \right) + A_3.$$

It follows that, if we choose c_1 with $c_1[K : \mathbb{Q}] > A_4 A_5$, then, for any sufficiently large d_1, d_2, d, we have

$$\frac{\max\limits_{i,a,b,c,a',b',c'} \log \|t_{i,abca'b'c'}\|_K}{[K : \mathbb{Q}]} \leq \frac{c_1}{\lambda_0}\left(1 + \frac{\log(d_1 + d_2 + d)}{d}\right)(d_1 + d_2 + d).$$

Note that $|t_{i,abca'b'c'}|_v \leq 1$ for any nonarchimedean place v, because $t_{i,abca'b'c'} \in O_K$. Further, for any archimedean place v, we have

$$\log |t_{i,abca'b'c'}|_v \leq \log \|t_{i,abca'b'c'}\|_K \leq \max\limits_{i,a,b,c,a',b',c'} \log \|t_{i,abca'b'c'}\|_K.$$

Thus, for a sequence $(F_0, \ldots, F_m)$ of polynomials given by $t_{i,abca'b'c'}$'s, we have

$$h(F_0, \ldots, F_m) \leq \frac{c_1}{\lambda_0}\left(1 + \frac{\log(d_1 + d_2 + d)}{d}\right)(d_1 + d_2 + d).$$

Since $(F_0, \ldots, F_m)$ is an element of U, by Proposition 5.11, $(F_0, \ldots, F_m)$ gives rise to a nonzero global section s of the Vojta divisor $V(d_1, d_2, d)$, i.e., $s \in H^0(C \times C, \mathcal{O}_{C \times C}(V(d_1, d_2, d))) \setminus \{0\}$. By the definition of $h(s)$ (see (5.12)), we have

$$h(s) \leq h(F_0, \ldots, F_m) \leq \frac{c_1}{\lambda_0}\left(1 + \frac{\log(d_1 + d_2 + d)}{d}\right)(d_1 + d_2 + d).$$

This completes the proof of Theorem 5.4.

5.4 Upper Bound of the Index (the Proof of Theorem 5.5)

In this section, using Roth's lemma, we give an upper bound for the index of a global section s. This part is generalized by Faltings as the product theorem [9, Theorem 3.1].

Recall from Subsection 5.2.1 that $\Phi_{N\theta} : C \to \mathbb{P}^n$ is a closed embedding satisfying Assumption 5.7 and that, for each $1 \leq i \leq n$, the morphism $\Phi_{N\theta}^{0i} : C \to \mathbb{P}^1$ is given by the composition of $\Phi_{N\theta}$ and the projection $\mathbb{P}^n \dashrightarrow \mathbb{P}^1$ given by $(X_0 : \cdots : X_n) \mapsto (X_0 : X_i)$.

We write $\pi_{ij} = \Phi_{N\theta}^{0i} \times \Phi_{N\theta}^{0j} : C \times C \to \mathbb{P}^1 \times \mathbb{P}^1$. By Proposition 4.30, for $F \in K[X, X']_{(\delta_1, \delta_2)}$, $\mathrm{Norm}_{\pi_{ij}}(F(x, x'))$ is a bihomogeneous polynomial in $K[X_0, X_i, X_0', X_j']$ of bidegree $(N^2\delta_1, N^2\delta_2)$. Further, by Proposition 4.30, 2, there is a constant M depending only on $C, N, \Phi_{N\theta}, i$ and j such that

$$h(\mathrm{Norm}_{\pi_{ij}}(F(x, x'))) \leq N^2 h(F) + M(\delta_1 + \delta_2 + 1). \qquad (5.20)$$

The constant M may depend on i and j, but if we take the maximum ranging over i and j, then we may assume that M is independent of i and j. Further, there is a finite subset Z_1 of $C(\overline{K})$ such that if $z \notin Z_1$, then, for all i ($1 \leq i \leq n$), $\Phi_{N\theta}^{0i} : C \to \mathbb{P}^1$ is étale over $\Phi_{N\theta}^{0i}(z)$. Adding points z with $x_0 = 0$ to Z_1, we may assume that $x_0(z) \neq 0$ for $z \notin Z_1$.

Let s be a global section determined by a sequence $(F_0, \ldots, F_m)$ of polynomials (see Proposition 5.11). If we choose $(F_0, \ldots, F_m)$ suitably, we may assume that $h(F_0, \ldots, F_m) \leq h(s) + 1/2$ (see (5.12)).

We set $E = \mathrm{div}(s)$. Then $(\pi_{ij})_*(E)$ is a divisor on $\mathbb{P}^1 \times \mathbb{P}^1$ whose bidegree is (Nd_1, Nd_2). Indeed, the projection formula implies that

$$
\begin{aligned}
((\pi_{ij})_*(E) \cdot \mathcal{O}_{\mathbb{P}^1 \times \mathbb{P}^1}(0,1)) &= (E \cdot (\pi_{ij})^*(\mathcal{O}_{\mathbb{P}^1 \times \mathbb{P}^1}(0,1))) \\
&= \big(((d_1 - d)p_1^*(\theta) + (d_2 - d)p_2^*(\theta) + d\Delta) \cdot Np_2^*(\theta)\big) = Nd_1,
\end{aligned}
$$

and similarly $((\pi_{ij})_*(E) \cdot \mathcal{O}_{\mathbb{P}^1 \times \mathbb{P}^1}(1,0)) = Nd_2$. Thus, $\mathrm{Norm}_{\pi_{ij}}(s)$ is a bihomogeneous polynomial of bidegree (Nd_1, Nd_2), and we can consider the height of $\mathrm{Norm}_{\pi_{ij}}(s)$ as a polynomial. Further, since $s = F_0/y_0^d$, by Proposition 4.27, 1,

$$
\mathrm{Norm}_{\pi_{ij}}(F_0(\boldsymbol{x}, \boldsymbol{x}')) = \mathrm{Norm}_{\pi_{ij}}(s) \, \mathrm{Norm}_{\pi_{ij}}(y_0^d).
$$

Since $\deg(\mathrm{Norm}_{\pi_{ij}}(F_0(\boldsymbol{x}, \boldsymbol{x}'))) = N^2(\delta_1 + \delta_2)$ and $h(\mathrm{Norm}_{\pi_{ij}}(y_0^d)) \geq 0$, it follows from Proposition 4.8, 2 and the above estimate (5.20) that

$$
\begin{aligned}
h(\mathrm{Norm}_{\pi_{ij}}(s)) &\leq h(\mathrm{Norm}_{\pi_{ij}}(s)) + h(\mathrm{Norm}_{\pi_{ij}}(y_0^d)) \\
&\leq h(\mathrm{Norm}_{\pi_{ij}}(F_0(\boldsymbol{x}, \boldsymbol{x}'))) + 4N^2(\delta_1 + \delta_2) \\
&\leq N^2 h(F_0) + (M + 4N^2)(\delta_1 + \delta_2 + 1) \\
&\leq N^2 h(F_0, \ldots, F_m) + (M + 4N^2)(\delta_1 + \delta_2 + 1) \\
&\leq N^2 h(s) + N^2/2 + (M + 4N^2)(\delta_1 + \delta_2 + 1).
\end{aligned}
$$

Note that $\delta_1 = (d_1 + td)/N$ and $\delta_2 = (d_2 + td)/N$, so there is a constant M' depending only on C, N, t, and $\Phi_{N\theta}$ such that

$$
h(\mathrm{Norm}_{\pi_{ij}}(s)) + 4Nd_1 \leq N^2 h(s) + M'(d_1 + d_2 + d) \tag{5.21}
$$

for any i and j and any nonzero section s. Here, we have added $4Nd_1$ on the left-hand side for a later purpose.

Further, by Corollary 3.37, 2, there is a constant A depending only on the choice of the basis $\{\phi_0, \ldots, \phi_n\}$ of $H^0(C, \mathcal{O}_C(N\theta))$ such that

$$
\frac{N}{2g}|z|^2 \leq h_{N\theta}(z) + A \tag{5.22}
$$

for all $z \in C(\overline{K})$.

With the above setting, we put

$$Z = Z_1, \quad c_2 = 2ng, \quad c_3 = \frac{2gn}{N^2}M' + \frac{2g}{N}A, \quad \text{and} \quad c_4 = 4N. \quad (5.23)$$

We are going to prove Theorem 5.5 with these Z and c_2, c_3, c_4.

Let $z, z' \in C(\overline{K}) \setminus Z$. By Lemma 3.12,3,

$$h_{N\theta}(z) \le \sum_{\nu=1}^{n} h^+((x_\nu/x_0)(z)).$$

We take i such that

$$h^+((x_i/x_0)(z)) = \max_{1 \le \nu \le n} \left\{ h^+((x_\nu/x_0)(z)) \right\}.$$

Similarly, we take j such that

$$h^+((x_j/x_0)(z')) = \max_{1 \le \nu \le n} \left\{ h^+((x_\nu/x_0)(z')) \right\}.$$

Then

$$h_{N\theta}(z) \le nh^+((x_i/x_0)(z)) \quad \text{and} \quad h_{N\theta}(z') \le nh^+((x_j/x_0)(z')). \quad (5.24)$$

Let P be the polynomial given by $P(x, x') = \mathrm{Norm}_{\pi_{ij}}(s)(1, x, 1, x')$, and let $a = (x_i/x_0)(z)$ and $b = (x_j/x_0)(z')$. Note that P is a polynomial of bidegree at most (Nd_1, Nd_2). We are going to apply Roth's lemma (Theorem 4.20) to P and (a, b). By Proposition 4.27,2, $\pi_{ij}^*(\mathrm{Norm}_{\pi_{ij}}(s))$ is divisible by s, so

$$\mathrm{ind}_{(z,z')}(s; (d_1, d_2)) \le \mathrm{ind}_{(z,z')}(\pi_{ij}^*(\mathrm{Norm}_{\pi_{ij}}(s)); (d_1, d_2)).$$

Further, since $z, z' \notin Z$, the completion of the local ring at (z, z') is isomorphic to the completion of the local ring at (a, b) over $\overline{\mathbb{Q}}$, so

$$\mathrm{ind}_{(z,z')}(\pi_{ij}^*(\mathrm{Norm}_{\pi_{ij}}(s)); (d_1, d_2)) = \mathrm{ind}_{(a,b)}(P; (d_1, d_2))$$
$$= N\,\mathrm{ind}_{(a,b)}(P; (Nd_1, Nd_2)),$$

and thus, $\mathrm{ind}_{(z,z')}(s; (d_1, d_2)) \le N\,\mathrm{ind}_{(a,b)}(P; (Nd_1, Nd_2))$. In order to apply Roth's lemma, we need to verify

$$h(P) + 4Nd_1 \le \epsilon^2 \min\{Nd_1 h^+(a), Nd_2 h^+(b)\}.$$

Indeed,

$h(P) + 4Nd_1$

$$\le N^2 h(s) + M'(d_1 + d_2 + d) \qquad \text{(by (5.21))}$$

$$\le \frac{N}{n}\left(\epsilon^2 \min\left\{ d_1 \frac{N}{2g}|z|^2, d_2 \frac{N}{2g}|z'|^2 \right\} - A(d_1 + d_2 + d) \right)$$

$$\text{(by the assumption in Theorem 5.5 and (5.23))}$$

$$\leq \frac{N}{n}\left(\epsilon^2 \min\{d_1 h_{N\theta}(z), d_2 h_{N\theta}(z)\} - (1 - \epsilon^2)A(d_1 + d_2) - Ad\right)$$

(by (5.22))

$$\leq \epsilon^2 \frac{N}{n} \min\{d_1 h_{N\theta}(z), d_1 h_{N\theta}(z)\} \qquad \text{(by } \epsilon < 1\text{)}$$

$$\leq \epsilon^2 \min\{N d_1 h^+(a), N d_2 h^+(b)\} \qquad \text{(by (5.24))}.$$

Then we apply Roth's lemma (Theorem 4.20), and we obtain

$$\mathrm{ind}_{(a,b)}(P; (Nd_1, Nd_2)) \leq 4\epsilon,$$

which implies the conclusion of Theorem 5.5.

5.5 Lower Bound of the Index (the Proof of Theorem 5.6)

In this section, we give a lower bound of the index (Theorem 5.6). The computation is involved, but it is a simple application of the local Eisenstein theorem (see Lemma 4.31).

We begin with the following claim:

Claim 3 For any $0 \leq \nu \leq n$ and $0 \leq \mu \leq n$ with $\nu \neq \mu$, there exists a polynomial $P_{\nu,\mu}(U, V) \in K[U, V]$ with degree at most $2N$ such that $P_{\nu,\mu}(x_1/x_0, x_\nu/x_\mu) = 0$ and such that $(\partial P_{\nu,\mu}/\partial V)(x_1/x_0, x_\nu/x_\mu)$ is not identically zero. Further, for any $z \in C(K)$ with $x_0(z) \neq 0$, if we set $P'_{\nu,\mu}(U, V) = P_{\nu,\mu}(U + (x_1/x_0)(z), V)$, then we have the following:

1. $h^+(P'_{\nu,\mu}) \leq h^+(P_{\nu,\mu}) + \log(2N + 1) + 2N \log(2) + 2N h_{N\theta}(z)$.
2. If $x_\mu(z) \neq 0$, then

$$h^+((\partial P'_{\nu,\mu}/\partial V)(0, (x_\nu/x_\mu)(z))) \leq h^+(P'_{\nu,\mu}) + 2\log(2N) + (2N - 1)h_{N\theta}(z).$$

Proof Let $\pi : C \to \mathbb{P}^1$ be the morphism defined by $w \mapsto (x_0(w) : x_1(w))$, and $\pi' : C \to \mathbb{P}^1$ the morphism by $w \mapsto (x_\nu(w) : x_\mu(w))$. We set $\Pi = \pi \times \pi' : C \to \mathbb{P}^1 \times \mathbb{P}^1$.

Let $\deg(\Pi)$ denote the degree of $C \to \Pi(C)$. Then $\Pi(C)$ is geometrically irreducible[3] and with bidegree $(N/\deg(\Pi), N/\deg(\Pi))$ in $\mathbb{P}^1 \times \mathbb{P}^1$.

---

[3] The image of a geometrically irreducible, algebraic variety over a field of characteristic 0 is geometrically irreducible. This follows from the following fact: Let $k \subseteq F \subseteq E$ be extensions of fields, and let $\bar{k}$ be an algebraic closure of k. Then, if $E \otimes_k \bar{k}$ is an integral domain, then so is $F \otimes_k \bar{k}$. (Indeed, since $\bar{k}$ is flat over k, $F \otimes_k \bar{k} \to E \otimes_k \bar{k}$ is injective.)

Indeed, let (a, b) be the bidegree of $\Pi(C)$, and let $p_i : \mathbb{P}^1 \times \mathbb{P}^1 \to \mathbb{P}^1$ be the projection to the i-th factor. Then

$$
\begin{aligned}
N &= \deg(\pi'^*(\mathcal{O}_{\mathbb{P}^1}(1))) = \deg(\Pi^* p_2^*(\mathcal{O}_{\mathbb{P}^1}(1))) \\
&= \deg(\Pi)\deg(p_2^*(\mathcal{O}_{\mathbb{P}^1}(1))|_{\Pi(C)}) = \deg(\Pi)\left(\mathcal{O}_{\mathbb{P}^1 \times \mathbb{P}^1}(0, 1) \cdot \Pi(C)\right) = \deg(\Pi)a,
\end{aligned}
$$

so we obtain $a = N/\deg(\Pi)$. Similarly, $b = N/\deg(\Pi)$.

Let $P_{v,\mu} \in K[U, V]$ be a defining equation of $\Pi(C)$ on $X_1 \neq 0$ and $X_\mu \neq 0$. Then $P_{v,\mu}$ is geometrically irreducible and with degree at most $2N$, and satisfies $P_{v,\mu}(x_1/x_0, x_v/x_\mu) = 0$. Since the characteristic of K is zero, $p_1 : \Pi(C) \to \mathbb{P}^1$ is étale over a nonempty open set of $\mathbb{P}^1$. Hence, $(\partial P_{v,\mu}/\partial V)(x_1/x_0, x_v/x_\mu)$ is not identically zero.

We are going to show the inequality 1. We write

$$
P_{v,\mu} = \sum_{i, j} p_{ij} U^i V^j \qquad (p_{ij} \in K).
$$

Then

$$
P'_{v,\mu} = \sum_{i, j} p_{ij}(U + (x_1/x_0)(z))^i V^j = \sum_{l, j}\left(\sum_{i \geq l} \binom{i}{l} p_{ij}(x_1/x_0)^{i-l}(z)\right) U^l V^j.
$$

It follows that, if v is nonarchimedean, then

$$
|P'_{v,\mu}|_v \leq \max_{i, j}\{|p_{ij}|_v\} \max\{1, |(x_1/x_0)(z)|_v\}^{2N}
$$

$$
\leq |P_{v,\mu}|_v \max\{1, |(x_1/x_0)(z)|_v\}^{2N}.
$$

In addition, if v is archimedean, then

$$
|P'_{v,\mu}|_v \leq (2N + 1) \max_{i, j}\{|p_{ij}|_v\} \cdot 2^{2N} \cdot \max\{1, |(x_1/x_0)(z)|_v\}^{2N}
$$

$$
\leq (2N + 1)|P_{v,\mu}|_v 2^{2N} \max\{1, |(x_1/x_0)(z)|_v\}^{2N}.
$$

In summary, we obtain

$$
h^+(P'_{v,\mu}) \leq h^+(P_{v,\mu}) + \log(2N + 1) + 2N\log(2) + 2Nh^+((x_1/x_0)(z)).
$$

Further, since

$$
h^+((x_1/x_0)(z)) = h(x_0(z) : x_1(z))
$$

$$
\leq h(x_0(z) : x_1(z) : \cdots : x_n(z)) = h_{N\theta}(z),
$$

we obtain 1.

Lastly, let us show the inequality 2. We write $P'_{v,\mu} = \sum_{ij} p'_{ij} U^i V^j$ ($p'_{ij} \in K$). Then

$$(\partial P'_{v,\mu}/\partial V)(0, (x_v/x_\mu)(z)) = \sum_{j=1}^{2N} j p'_{0j} (x_v/x_\mu)^{j-1}(z).$$

It follows that, if v is nonarchimedean, then

$$|(\partial P'_{v,\mu}/\partial V)(0, (x_v/x_\mu)(z))|_v \leq \max_{i,j} |p'_{ij}|_v \max_{0 \leq j \leq 2N-1} \{|(x_v/x_\mu)^j (z)|_v\}$$

$$\leq \max_{i,j} |p'_{ij}|_v \max\{1, |(x_v/x_\mu)(z)|_v\}^{2N-1},$$

so we get

$$\log^+ (|(\partial P'_{v,\mu}/\partial V)(0, (x_v/x_\mu)(z))|_v)$$

$$\leq \max_{i,j} \log^+ (|p'_{ij}|_v) + (2N-1)\log^+ \left(|(x_v/x_\mu)(z)|_v\right).$$

If v is archimedean, then

$$|(\partial P'_{v,\mu}/\partial V)(0, (x_v/x_\mu)(z))|_v$$

$$\leq (2N)^2 \max_{i,j} |p'_{ij}|_v \max_{0 \leq j \leq 2N-1} \{|(x_v/x_\mu)^j (z)|_v\}$$

$$\leq (2N)^2 \max_{i,j} |p'_{ij}|_v \max\{1, |(x_v/x_\mu)(z)|_v\}^{2N-1},$$

so we get

$$\log^+ (|(\partial P'_{v,\mu}/\partial V)(0, (x_v/x_\mu)(z))|_v)$$

$$\leq 2\log(2N) + \max_{i,j} \log^+ (|p'_{ij}|_v) + (2N-1)\log^+ \left(|(x_v/x_\mu)(z)|_v\right).$$

In summary, we obtain

$$h^+ ((\partial P'_{v,\mu}/\partial V)(0, (x_v/x_\mu)(z)))$$

$$\leq 2\log(2N) + h^+(P'_{v,\mu}) + (2N-1)h^+((x_v/x_\mu)(z))$$

$$\leq 2\log(2N) + h^+(P'_{v,\mu}) + (2N-1)h_{N\theta}(z),$$

as required.   □

As in the proof of Claim 3, let π denote the projection $C \to \mathbb{P}^1$ defined by $w \mapsto (x_0(w) : x_1(w))$. If we set $T = x_1/x_0$, then T is a coordinate function of $\mathbb{P}^1 \setminus \{\infty\}$. We put $Z' = \{\infty\} \cup \{\zeta \in \mathbb{P}^1 \mid \pi \text{ is not étale over } \zeta\}$ and

$$Z_2 = \pi^{-1}(Z') \cup \bigcup_j \{w \in C(\overline{K}) \mid x_j(w) = 0\} \cup$$

$$\times \bigcup_{v \neq \mu} \{w \in C(\overline{K}) \mid (\partial P_{v,\mu}/\partial V)((x_1/x_0)(w), (x_v/x_\mu)(w)) = 0\}$$

as the exceptional set in Theorem 5.6. In what follows, we assume that $z, z' \in C(K) \setminus Z_2$. If we set $t = T - x_1(z)/x_0(z)$, then t gives a local coordinate function around z. Similarly, $t' = T - x_1(z')/x_0(z')$ gives a local coordinate function around z'. We put

$$\partial_i = \frac{1}{i!} \left(\frac{\partial}{\partial t} \right)^i \quad \text{and} \quad \partial'_{i'} = \frac{1}{i'!} \left(\frac{\partial}{\partial t'} \right)^{i'}.$$

Let $s \in H^0\big(C \times C, \mathcal{O}_{C \times C}(V(d_1, d_2, d))\big)$ be a section represented by a sequence of polynomials $(F_0(X, X'), \ldots, F_m(X, X'))$. We are going to estimate a lower bound of the index $\mathrm{ind}_{(z, z')}(s; (d_1, d_2))$.

We first recall the height $h_{V(d_1, d_2, d)} : (C \times C)(\overline{K}) \to \mathbb{R}$ associated to the Vojta divisor $V(d_1, d_2, d)$. The Vojta divisor is defined by $V(d_1, d_2, d) = \delta_1 p_1^*(N\theta) + \delta_2 p_2^*(N\theta) - dB$ in (5.8). Recall that

$$\Phi_{N\theta} : C \to \mathbb{P}^n, \quad w \mapsto (x_0(w) : \cdots : x_n(w)),$$

$$\Phi_{|B|} : C \times C \to \mathbb{P}^m, \quad (w, w') \mapsto (y_0(w, w') : \cdots : y_m(w, w')).$$

Then the height function $h_{V(d_1, d_2, d)}$ associated to the Vojta divisor $V(d_1, d_2, d)$ on $C \times C$ is defined by

$$h_{V(d_1, d_2, d)}(w, w') = \delta_1 h_{N\theta}(w) + \delta_2 h_{N\theta}(w') - d h_B(w, w'),$$

where h is the absolute (logarithmic) Weil height on projective space and

$$\begin{cases} h_{N\theta}(w) = h(x_0(w) : \cdots : x_n(w)), \\ h_B(w, w') = h\big(y_0(w, w') : \cdots : y_m(w, w')\big). \end{cases}$$

We consider $h_{V(d_1, d_2, d)}(z, z')$. We set $a_j = x_j(z)$, $a'_{j'} = x_{j'}(z')$, $b_i = y_i(z, z')$. From the assumption that $z, z' \notin Z_2$, we have $a_j \neq 0$ and $a'_{j'} \neq 0$. For each absolute value $v \in M_K$, we choose $j(v), j'(v), i(v)$ such that $|a_{j(v)}|_v = \max_j\{|a_j|_v\}$, $|a'_{j'(v)}|_v = \max_{j'}\{|a'_{j'}|_v\}$, $|b_{i(v)}|_v = \max_i\{|b_i|_v\}$. We note that $b_{i(v)} \neq 0$. Then

$$[K : \mathbb{Q}]h_{V(d_1,d_2,d)}(z,z')$$
$$= \delta_1 \sum_v \max_j \log(|a_j|_v) + \delta_2 \sum_v \max_{j'} \log(|a'_{j'}|_v) - d \sum_v \max_i \log(|b_i|_v)$$
$$= \sum_v \log \left| \frac{a_{j(v)}^{\delta_1} a_{j'(v)}'^{\delta_2}}{b_{i(v)}^d} \right|_v. \tag{5.25}$$

We fix τ with $y_\tau(z,z') \neq 0$, and we define the rational function g by

$$g = \frac{s}{x_0^{\delta_1} x_0'^{\delta_2} y_\tau^{-d}} \left(= \frac{F_i(\boldsymbol{x}/x_0, \boldsymbol{x}'/x_0')}{(y_i/y_\tau)^d} \right).$$

We take (l,l') such that

$$\mathrm{ind}_{(z,z')}(s; (d_1,d_2)) = \frac{l}{d_1} + \frac{l'}{d_2}.$$

It follows from the definition of the index that, for any $(\alpha,\alpha') \in \mathbb{Z}_{\geq 0}^2$ with $\alpha \leq l, \alpha' \leq l', (\alpha,\alpha') \neq (l,l')$, we have

$$(\partial_\alpha \partial'_{\alpha'} g)(z,z') = 0. \tag{5.26}$$

We set $f = (\partial_l \partial'_{l'} g)(z,z')$. We claim the following:

Claim 4 $f \neq 0$, and for any i, j, j', we have

$$f = (b_i/b_\tau)^{-d} (a_j/a_0)^{\delta_1} (a'_{j'}/a'_0)^{\delta_2} \partial_l \partial'_{l'} F_i(\boldsymbol{x}/x_j, \boldsymbol{x}'/x'_{j'})(z,z').$$

Proof By the definition of the index, we have $f \neq 0$. On the other hand, by the definition of g, we have

$$g = (y_i/y_\tau)^{-d} F_i(\boldsymbol{x}/x_0, \boldsymbol{x}'/x_0')$$
$$= (y_i/y_\tau)^{-d} (x_j/x_0)^{\delta_1} (x'_{j'}/x_0')^{\delta_2} F_i(\boldsymbol{x}/x_j, \boldsymbol{x}'/x'_{j'}).$$

In other words,

$$F_i(\boldsymbol{x}/x_j, \boldsymbol{x}'/x'_{j'}) = (y_i/y_\tau)^d (x_j/x_0)^{-\delta_1} (x'_{j'}/x_0')^{-\delta_2} g.$$

By the Leibniz rule, we have

$$\partial_l \partial'_{l'}(F_i(\boldsymbol{x}/x_j, \boldsymbol{x}'/x'_{j'}))$$
$$= \sum_{\substack{e_1+e_2=l \\ e'_1+e'_2=l'}} \partial_{e_1} \partial'_{e'_1}((y_i/y_\tau)^d (x_j/x_0)^{-\delta_1} (x'_{j'}/x_0')^{-\delta_2}) \partial_{e_2} \partial'_{e'_2}(g),$$

so, by using (5.26), we obtain

$$\partial_l \partial'_{l'}(F_i(\boldsymbol{x}/x_j, \boldsymbol{x}'/x'_{j'}))(z, z') = (b_i/b_\tau)^d (a_j/a_0)^{-\delta_1} (a'_{j'}/a'_0)^{-\delta_2} (\partial_l \partial'_{l'} g)(z, z')$$
$$= (b_i/b_\tau)^d (a_j/a_0)^{-\delta_1} (a'_{j'}/a'_0)^{-\delta_2} f,$$

as required. $\qquad\qquad\qquad\qquad\qquad\qquad\qquad\qquad\qquad\qquad\qquad\qquad$ $\square$

We return to (5.25):

$$[K : \mathbb{Q}] h_{V(d_1, d_2, d)}(z, z') = \sum_v \log \left| \frac{a_{j(v)}{}^{\delta_1} a'_{j'(v)}{}^{\delta_2}}{b_{i(v)}^d} \right|_v.$$

By the product formula and Claim 4, we rewrite the above equality as

$$[K : \mathbb{Q}] h_{V(d_1, d_2, d)}(z, z') = \sum_v \log \left| \frac{a_{j(v)}{}^{\delta_1} a'_{j'(v)}{}^{\delta_2}}{f b_{i(v)}^d} \right|_v$$
$$= -\sum_v \log \left| \partial_l \partial'_{l'} F_{i(v)}(\boldsymbol{x}/x_{j(v)}, \boldsymbol{x}'/x'_{j'(v)})(z, z') \frac{b_\tau^d}{a_0^{\delta_1} a'_0{}^{\delta_2}} \right|_v$$
$$= -\sum_v \log \left| \partial_l \partial'_{l'} F_{i(v)}(\boldsymbol{x}/x_{j(v)}, \boldsymbol{x}'/x'_{j'(v)})(z, z') \right|_v.$$

We set

$$F_i(X, X') = \sum_{\substack{I, I' \in \mathbb{Z}_{\geq 0}^{n+1} \\ |I| = \delta_1, |I'| = \delta_2}} p_{i, I, I'} X^I X'^{I'}.$$

Then

$$\partial_l \partial'_{l'} F_{i(v)}(\boldsymbol{x}/x_{j(v)}, \boldsymbol{x}'/x'_{j'(v)})(z, z')$$
$$= \sum_{\substack{I, I' \in \mathbb{Z}_{\geq 0}^{n+1} \\ |I| = \delta_1, |I'| = \delta_2}} p_{i(v), I, I'} \partial_l ((\boldsymbol{x}/x_{j(v)})^I)(z) \partial'_{l'} ((\boldsymbol{x}'/x'_{j'(v)})^{I'})(z').$$

If v is nonarchimedean, then

$$|\partial_l \partial'_{l'} F_{i(v)}(\boldsymbol{x}/x_{j(v)}, \boldsymbol{x}'/x'_{j'(v)})(z, z')|_v$$
$$\leq \max_{I, I'} \left\{ |p_{i(v), I, I'} \partial_l ((\boldsymbol{x}/x_{j(v)})^I)(z) \partial'_{l'} ((\boldsymbol{x}'/x'_{j'(v)})^{I'})(z')|_v \right\}$$
$$\leq \max_{i, I, I'} \left\{ |p_{i, I, I'}|_v \right\} \max_I \left\{ |\partial_l ((\boldsymbol{x}/x_{j(v)})^I)(z)|_v \right\} \max_{I'} \left\{ |\partial'_{l'} ((\boldsymbol{x}'/x'_{j'(v)})^{I'})(z')|_v \right\}.$$

If v is archimedean, then

$$|\partial_l \partial_{l'}' F_{i(v)}(\boldsymbol{x}/x_{j(v)}, \boldsymbol{x}'/x'_{j'(v)})(z, z')|_v$$

$$\leq \binom{\delta_1 + n}{n}\binom{\delta_2 + n}{n} \times$$

$$\max_{I, I'}\left\{|P_{i(v), I, I'}\partial_l((\boldsymbol{x}/x_{j(v)})^I)(z)\partial_{l'}'((\boldsymbol{x}'/x'_{j'(v)})^{I'})(z')|_v\right\}$$

$$\leq \binom{\delta_1 + n}{n}\binom{\delta_2 + n}{n} \max_{i, I, I'}\left\{|P_{i, I, I'}|_v\right\} \times$$

$$\max_{I}\left\{|\partial_l((\boldsymbol{x}/x_{j(v)})^I)(z)|_v\right\}\max_{I'}\left\{|\partial_{l'}'((\boldsymbol{x}'/x'_{j'(v)})^{I'})(z')|_v\right\}.$$

Thus, noting that $\binom{\delta_1+n}{n}\binom{\delta_2+n}{n} \leq 2^{\delta_1+n}2^{\delta_2+n}$, we have

$$[K : \mathbb{Q}]h_{V(d_1, d_2, d)}(z, z')$$

$$\geq -[K : \mathbb{Q}]h(F_0, \ldots, F_m) - [K : \mathbb{Q}]\log(2)(\delta_1 + \delta_2 + 2n)$$

$$- \sum_v \max_{\substack{I \in \mathbb{Z}_{\geq 0}^{n+1} \\ |I|=\delta_1}}\left\{\log^+ |\partial_l((\boldsymbol{x}/x_{j(v)})^I)(z)|_v\right\}$$

$$- \sum_v \max_{\substack{I' \in \mathbb{Z}_{\geq 0}^{n+1} \\ |I'|=\delta_2}}\left\{\log^+ |\partial_{l'}'((\boldsymbol{x}'/x'_{j'(v)})^{I'})(z')|_v\right\}. \quad (5.27)$$

Claim 5   There exists a positive constant A_1 that depends only on $C, N, \Phi_{N\theta}$, and $h^+(P_{v,\mu})$ with $1 \leq v \leq n$, $1 \leq \mu \leq n$, $v \neq \mu$ such that

$$\sum_v \max_{\substack{I \in \mathbb{Z}_{\geq 0}^{n+1} \\ |I|=\delta_1}}\left\{\log^+ |\partial_l((\boldsymbol{x}/x_{j(v)})^I)(z)|_v\right\} \leq [K : \mathbb{Q}](\delta_1 \log(2) + A_1(1 + |z|^2)l)$$

and

$$\sum_v \max_{\substack{I' \in \mathbb{Z}_{\geq 0}^{n+1} \\ |I'|=\delta_2}}\left\{\log^+ |\partial_{l'}'((\boldsymbol{x}'/x'_{j'(v)})^{I'})(z')|_v\right\} \leq [K : \mathbb{Q}](\delta_2 \log(2) + A_1(1 + |z'|^2)l').$$

Proof   For $I = (e_0, \ldots, e_n) \in \mathbb{Z}_{\geq 0}^{n+1}$, we set $I(j) = e_j$. We define $J \in \mathbb{Z}_{\geq 0}^{n+1}$ by

$$J(j) = \begin{cases} \delta_1 & (\text{if } j = j(v)), \\ 0 & (\text{if } j \neq j(v)). \end{cases}$$

Since $(\boldsymbol{x}/x_{j(v)})^J = 1$, we have

$$\sum_{v} \max_{\substack{I \in \mathbb{Z}_{\geq 0}^{n+1} \\ |I| = \delta_1}} \left\{ \log^+ |\partial_l((\boldsymbol{x}/x_{j(v)})^I)(z)|_v \right\}$$

$$= \sum_{v} \max_{\substack{I \in \mathbb{Z}_{\geq 0}^{n+1} \setminus \{J\} \\ |I| = \delta_1}} \left\{ \log^+ |\partial_l((\boldsymbol{x}/x_{j(v)})^I)(z)|_v \right\}.$$

For $I \in \mathbb{Z}_{\geq 0}^{n+1} \setminus \{J\}$ with $|I| = \delta_1$, we are going to estimate

$$\log^+(|\partial_l((\boldsymbol{x}/x_{j(v)})^I)(z)|_v).$$

We set

$$P'_{v,\mu}(X,Y) = P_{v,\mu}(X + (x_0/x_1)(z), Y).$$

We will use the local Eisenstein theorem (Corollary 4.32). Since

$$(\boldsymbol{x}/x_{j(v)})^I = (x_0/x_{j(v)})^{I(0)} \cdots (x_{j(v)-1}/x_{j(v)})^{I(j(v)-1)}$$
$$\times (x_{j(v)+1}/x_{j(v)})^{I(j(v)+1)} \cdots (x_n/x_{j(v)})^{I(n)},$$

the e in Corollary 4.32 equals $|I| - I(j(v))$. By the definition of $j(v)$, we have $|(x_v/x_{j(v)})(z)|_v \leq 1$, and by the definition of the exceptional set Z_2, we have

$$(\partial P'_{v,j(v)}/\partial Y)(0, (x_v/x_{j(v)})(z)) \neq 0.$$

We, thus, obtain the following from Corollary 4.32:
If v is nonarchimedean, then

$$\log^+(|\partial_l((\boldsymbol{x}/x_{j(v)})^I)(z)|_v)$$

$$\leq 2l \max_{v \neq j(v)} \left\{ \log^+ \left(\frac{|P'_{v,j(v)}|_v}{|(\partial P'_{v,j(v)}/\partial Y)(0, (x_v/x_{j(v)})(z))|_v} \right) \right\}.$$

If v is archimedean, then, noting that

$$\binom{l + |I| - I(j(v)) - 1}{l} \leq 2^{l+|I|-I(j(v))-1} \leq 2^{l+\delta_1},$$

we have

$$\log^+(|\partial_l((\boldsymbol{x}/x_{j(v)})^I)(z)|_v) \leq (l + \delta_1) \log(2) + 7l \log(4N)$$

$$+ 2l \max_{v \neq j(v)} \left\{ \log^+ \left(\frac{|P'_{v,j(v)}|_v}{|(\partial P'_{v,j(v)}/\partial Y)(0, (x_v/x_{j(v)})(z))|_v} \right) \right\}.$$

Taking the sum over all v, we obtain

$$\sum_v \max_{\substack{I \in \mathbb{Z}_{\geq 0}^{n+1} \\ |I|=\delta_1}} \left\{ \log^+ |\partial_I((x/x_{j(v)})^I)(z)|_v \right\}$$

$$\leq (\delta_1 \log(2) + l(\log(2) + 7\log(4N)))[K : \mathbb{Q}]$$

$$+ 2l \sum_v \max_{v \neq j(v)} \left\{ \log^+ \left(\frac{|P'_{v,j(v)}|_v}{|(\partial P'_{v,j(v)}/\partial Y)(0, (x_v/x_{j(v)})(z))|_v} \right) \right\}.$$

Further, let us estimate the last term of the above equation

$$H = \sum_v \max_{v \neq j(v)} \left\{ \log^+ \left(\frac{|P'_{v,j(v)}|_v}{|(\partial P'_{v,j(v)}/\partial Y)(0, (x_v/x_{j(v)})(z))|_v} \right) \right\}.$$

We obtain

$$H \leq \sum_v \max_{v \neq \mu} \left\{ \log^+ \left(\frac{|P'_{v,\mu}|_v}{|(\partial P'_{v,\mu}/\partial Y)(0, (x_v/x_\mu)(z))|_v} \right) \right\}$$

$$\leq \sum_v \sum_{v \neq \mu} \log^+ \left(\frac{|P'_{v,\mu}|_v}{|(\partial P'_{v,\mu}/\partial Y)(0, (x_v/x_\mu)(z))|_v} \right)$$

$$= \sum_{v \neq \mu} \sum_v \log^+ \left(\frac{|P'_{v,\mu}|_v}{|(\partial P'_{v,\mu}/\partial Y)(0, (x_v/x_\mu)(z))|_v} \right)$$

$$\leq \sum_{v \neq \mu} \sum_v \left\{ \log^+(|P'_{v,\mu}|_v) + \log^+ \left(\frac{1}{|(\partial P'_{v,\mu}/\partial Y)(0, (x_v/x_\mu)(z))|_v} \right) \right\}$$

$$= [K : \mathbb{Q}] \sum_{v \neq \mu} \left(h^+(P'_{v,\mu}) + h^+ \left(\frac{1}{(\partial P'_{v,\mu}/\partial Y)(0, (x_v/x_\mu)(z))} \right) \right)$$

$$= [K : \mathbb{Q}] \sum_{v \neq \mu} \left(h^+(P'_{v,\mu}) + h^+ \left((\partial P'_{v,\mu}/\partial Y)(0, (x_v/x_\mu)(z)) \right) \right).$$

Here, for the last equality, we have used Proposition 3.5,4.

Thus, by Claim 3, there exist constants B_1, B_2 that depend on N, n, and $h^+(P_{v,\mu})$ with $1 \leq v \leq n$, $1 \leq \mu \leq n$, $v \neq \mu$ such that

$$H \leq [K : \mathbb{Q}](B_1 + B_2 h_{N\theta}(z)).$$

Replacing constants B_1, B_2 suitably, we get

$$\frac{1}{[K : \mathbb{Q}]} \sum_v \max_{\substack{I \in \mathbb{Z}_{\geq 0}^{n+1} \\ |I|=\delta_1}} \left\{ \log^+ |\partial_I((x/x_{j(v)})^I)(z)|_v \right\} \leq \delta_1 \log(2) + (B_1 + B_2 h_{N\theta}(z))l.$$

Further, by Corollary 3.37, 2, we have $h_{N\theta}(\cdot) = \frac{N}{2g}|\cdot|^2 + O(1)$. Hence, we obtain the first claim. The case for z' is similar. $\square$

By (5.27) and Claim 5, we have shown that

$$h_{V(d_1,d_2,d)}(z,z') \geq -h(F_0,\ldots,F_m) - 2\log(2)(\delta_1 + \delta_2 + n)$$
$$- A_1((1+|z|^2)l + (1+|z'|^2)l').$$

Thus, we obtain

$$h_{V(d_1,d_2,d)}(z,z') \geq -h(F_0,\ldots,F_m) - 2\log(2)(\delta_1 + \delta_2 + n)$$
$$- A_1\left(d_1(1+|z|^2)\frac{l}{d_1} + d_2(1+|z'|^2)\frac{l'}{d_2}\right)$$
$$\geq -h(F_0,\ldots,F_m) - 2\log(2)(\delta_1 + \delta_2 + n)$$
$$- A_1\max\{d_1(1+|z|^2), d_2(1+|z'|^2)\}\operatorname{ind}_{(z,z')}(s;(d_1,d_2)).$$

On the other hand, by Corollary 3.37, there exist positive constants α_1, α_2 such that, for any $w, w' \in C(\overline{K})$, we have

$$h_{N\theta}(w) \leq \frac{N}{2g}|w|^2 + \alpha_1, \quad h_B(w,w') \geq \frac{t}{2g}(|w|^2 + |w'|^2) + \langle w, w'\rangle - \alpha_2.$$

Thus, we have

$$h_{V(d_1,d_2,d)}(z,z')$$
$$\leq \delta_1\left(\frac{N}{2g}|z|^2 + \alpha_1\right) + \delta_2\left(\frac{N}{2g}|z'|^2 + \alpha_1\right)$$
$$- d\left(\frac{t}{2g}(|z|^2 + |z'|^2) + \langle z, z'\rangle - \alpha_2\right)$$
$$= \frac{d_1}{2g}|z|^2 + \frac{d_2}{2g}|z'|^2 - d\langle z, z'\rangle + \alpha_1(\delta_1 + \delta_2) + \alpha_2 d.$$

Now we take a constant A_2 that depends only on $\alpha_1, \alpha_2, t, N, n$ satisfying

$$\alpha_1(\delta_1 + \delta_2) + \alpha_2 d + 2\log(2)(\delta_1 + \delta_2 + n) \leq A_2(d_1 + d_2 + d).$$

(For example, we may take $A_2 = \frac{2t(\alpha_1 + 2\log(2))}{N} + 2n\log(2) + \alpha_2$.) Then we have

$$\operatorname{ind}_{(z,z')}(s;(d_1,d_2))$$
$$\geq \frac{d\langle z,z'\rangle - \frac{d_1}{2g}|z|^2 - \frac{d_2}{2g}|z'|^2 - h(F_0,\ldots,F_m) - A_2(d_1 + d_2 + d)}{A_1\max\{d_1(1+|z|^2), d_2(1+|z'|^2)\}}.$$

Here, we take the supremum of the right-hand side of the above equation under the conditions $(F_0, \ldots, F_m) \in U$ and $R(F_0, \ldots, F_m) = s$ (see (5.12)). We get

$$\mathrm{ind}_{(z,z')}(s;(d_1,d_2))$$
$$\geq \frac{2gd\langle z,z'\rangle - d_1|z|^2 - d_2|z'|^2 - 2g\,h(s) - 2gA_2(d_1 + d_2 + d)}{2gA_1 \max\{d_1(1 + |z|^2), d_2(1 + |z'|^2)\}}.$$

Then, if we set $c_5 = 2gA_2$, $c_6 = 2gA_1$, we obtain Theorem 5.6.

5.6 An Application to Fermat Curves

In the previous section, finally, we have completed the proof of Faltings's theorem. This section is a light dessert after the heavy main dish. We give an easy application of Faltings's theorem to Fermat curves.

Let K be a number field, and let $h \colon \mathbb{P}^n(K) \to \mathbb{R}$ be the absolute (logarithmic) Weil height defined in Section 3.3. Thus, the height of $x = (x_0 : \cdots : x_n) \in \mathbb{P}^n(K)$ is given by

$$h(x) = \frac{1}{[K : \mathbb{Q}]} \sum_{v \in M_K} \log \left(\max_{0 \leq i \leq n} \{|x_i|_v\} \right). \tag{5.28}$$

Let $(X : Y : Z)$ be the homogeneous coordinates of $\mathbb{P}^2$. Let n be a positive integer. The *Fermat* (Figure 5.5) *curve* F_n of degree n is the projective curve defined by

$$X^n + Y^n - Z^n = 0$$

Figure 5.5 Pierre de Fermat.
Source: https://en.wikipedia.org.

in $\mathbb{P}^2$. We say that F_n has the *Fermat property over K* if $F_n(K)$ consists of points with height 0, i.e., $F_n(K) \subseteq \{x \in \mathbb{P}^2(K) \mid h(x) = 0\}$.

Let μ_K be the set of roots of unity in K. It follows from Lemma 3.15 that if $h(x) = 0$ for $x = (x_0 : x_1 : x_2) \in \mathbb{P}^2(K)$, then there exist a $\lambda \in K^\times$ and $y_0, y_1, y_2 \in \mu_K \cup \{0\}$ such that $(x_0, x_1, x_2) = \lambda(y_0, y_1, y_2)$. Thus, F_n has the Fermat property over K if and only if the coordinates of any point in $F_n(K)$, after changing representatives, belong to $\mu_K \cup \{0\}$.

Suppose that $K = \mathbb{Q}$. Then Wiles's theorem (Fermat's last theorem) implies that, for any $n \geq 3$, F_n has the Fermat property over $\mathbb{Q}$. Before this great theorem, Granville and Heath-Brown had obtained a weaker result: They showed that, for almost all n, F_n has the Fermat property over $\mathbb{Q}$ (see [25] for details).

Now let K be a number field in general. The purpose of this section is to show that, for almost all n, F_n has the Fermat property over K, as an application of Faltings's theorem.

Over $\mathbb{Q}$, one of the coordinates of any point in $F_n(\mathbb{Q})$ is 0. We remark that this is not the case over a number field K in general. For example, let ω denote a third primitive root of unity, let n be a positive integer, and let x, y, z be roots of unity with $x^n = \omega^2$, $y^n = \omega$, $z^n = -1$. Since $\omega^2 + \omega = -1$, we see that $(x, y, z) \in F_n(\mathbb{Q}(x, y, z))$, but none of x, y, z is 0.

In the following, we fix a number field K. Let FP(K) be the set of all positive integers n such that F_n has the Fermat property over K.

Proposition 5.14 *We have*

$$\lim_{u \to \infty} \frac{\#(\text{FP}(K) \cap [1, u])}{u} = 1. \tag{5.29}$$

In other words, in the sense of Equation (5.29), for almost all n, F_n has the Fermat property over K.

First, we give the following key lemma: (The case $K = \mathbb{Q}$ is due to Filaseta. See [25] for details.)

Lemma 5.15 *If $n \geq 4$, then there exists a positive integer m_0 such that for any $m \geq m_0$, $mn \in$ FP(K).*

Proof Since $n \geq 4$, the genus of F_n is at least 2. By Faltings's theorem, $F_n(K)$ is a finite set. We set $H = \max_{x \in F_n(K)}\{h(x)\}$ and

$$a = \inf\{h(x) \mid x \in \mathbb{P}^2(K), \ h(x) > 0\}.$$

Since $h(\mathbb{P}^2(K))$ is discrete by Lemma 3.15, we have $a > 0$. We take a positive integer m_0 satisfying $m_0 \geq \exp(H/a)$. Let us show that, for any $m \geq m_0$, any element of $F_{mn}(K)$ has height 0. Otherwise, there is $x = (x_0 : x_1 : x_2) \in F_{mn}(K)$ with $h(x) > 0$. Then $h(x) \geq a$ and $y = (x_0^m : x_1^m : x_2^m) \in F_n(K)$. It follows that

$$H \geq h(y) = mh(x) \geq ma,$$

so $\exp(H/a) \geq \exp(m)$. Thus, $m \geq \exp(m)$, which is a contradiction. $\qquad\square$

To prove Proposition 5.14, it suffices to show the following lemma. Indeed, by Lemma 5.15, FP(K) satisfies the assumption of the following lemma, and we obtain Proposition 5.14.

Lemma 5.16 *Let Σ be a subset of $\mathbb{Z}_{>0}$. For each prime number $p \geq 5$, we assume that there exists m_0 such that $mp \in \Sigma$ for any $m \geq m_0$. Then we have*

$$\lim_{u \to \infty} \frac{\#(\Sigma \cap [1, u])}{u} = 1.$$

Proof We may assume that u is an integer. We put

$$T(u) = \#(\Sigma \cap [1, u]), \quad N(u) = \#((\mathbb{Z}_{>0} \setminus \Sigma) \cap [1, u]).$$

It follows from the Euler (Figure 5.6) product formula of the Riemann zeta function and $\zeta(1) = \infty$ that

$$\prod_{p \text{ is a prime}} (1 - 1/p) = 0.$$

Thus, for any $0 < \epsilon < 1$, there exist distinct prime numbers $p_1, \ldots, p_s$ such that $p_i \geq 5$ for each i and

$$(1 - 1/p_1) \cdots (1 - 1/p_s) \leq \epsilon/4.$$

Figure 5.6 Leonhard Euler.
Source: E. Handmann, Bildnis des Leonhard Euler, 1753 in: Kunstmuseum Basel, Sammlung Online.

From the assumption, for each p_i, there exists m_i such that $p_i m \in \Sigma$ for any $m \geq m_i$. We set $P = p_1 \cdots p_s$ and $M = \max_{1 \leq i \leq s}\{p_i m_i\}$. If n is an integer such that $M < n \leq u$ and $\mathrm{GCD}(n, P) \neq 1$, then $n \in \Sigma$. Thus,

$$N(u) \leq M + \#\{1 \leq n \leq u \mid \mathrm{GCD}(n, P) = 1\}.$$

Further, we have

$$\#\{1 \leq n \leq u \mid \mathrm{GCD}(n, P) = 1\} \leq (u/P + 1)\varphi(P)$$
$$= (u + P)(1 - 1/p_1) \cdots (1 - 1/p_s),$$

where

$$\varphi(P) = \#\{1 \leq n \leq P \mid \mathrm{GCD}(n, P) = 1\} = P(1 - 1/p_1) \cdots (1 - 1/p_s)$$

is Euler's function. Thus,

$$N(u) \leq M + (u + P)(1 - 1/p_1) \cdots (1 - 1/p_s).$$

We set $u_0 = \max\{2M/\epsilon, 2P/\epsilon\}$. Then, for any $u \geq u_0$, we have

$$\frac{N(u)}{u} \leq \frac{M}{u} + \left(1 + \frac{P}{u}\right)(1 - 1/p_1) \cdots (1 - 1/p_s)$$

$$\leq \frac{\epsilon}{2} + \left(1 + \frac{\epsilon}{2}\right)\frac{\epsilon}{4} \leq \frac{\epsilon}{2} + \frac{\epsilon}{2} = \epsilon.$$

It follows that $\lim_{u \to \infty} N(u)/u = 0$, so

$$\lim_{u \to \infty} \frac{T(u)}{u} = 1,$$

as required. $\qquad\qquad\qquad\qquad\qquad\qquad\qquad\qquad\qquad\qquad\qquad\qquad$ □

References

[1] Lars V. Ahlfors. *Complex analysis: An introduction to the theory of analytic functions of one complex variable*, 3rd ed. International Series in Pure and Applied Mathematics. New York: McGraw-Hill Book Co., 1978.

[2] E. Arbarello, M. Cornalba, P. A. Griffiths, and J. Harris. *Geometry of algebraic curves: Volume 1*, vol. 267 of Grundlehren der mathematischen Wissenschaften [Fundamental Principles of Mathematical Sciences]. New York: Springer-Verlag, 1985.

[3] M. F. Atiyah and I. G. Macdonald. *Introduction to commutative algebra*. Reading, MA, London, Ontario: Addison-Wesley Publishing Co., 1969.

[4] Christina Birkenhake and Herbert Lange. *Complex abelian varieties*, 2nd ed., Vol. 302 of Grundlehren der mathematischen Wissenschaften [Fundamental Principles of Mathematical Sciences]. Berlin: Springer-Verlag, 2004.

[5] Enrico Bombieri. The Mordell conjecture revisited. *Ann. Scuola Norm. Sup. Pisa Cl. Sci. (4)*, 17(4), 1990: 615–40.

[6] Enrico Bombieri and Walter Gubler. *Heights in Diophantine geometry*, Vol. 4 of *New Mathematical Monographs*. Cambridge: Cambridge University Press, 2006.

[7] B. Edixhoven and J.-H. Evertse, eds. *Diophantine approximation and abelian varieties*, Vol. 1566 of Lecture Notes in Mathematics. Berlin: Springer-Verlag, 1993. Introductory lectures, Papers from the conference held in Soesterberg, April 12–16, 1992.

[8] Gerd Faltings. Finiteness theorems for abelian varieties over number fields. In Gary Cornell and Joseph H. Silverman (eds.), *Arithmetic geometry (Storrs, Conn., 1984)* Trans. Edward Shipz. [Invent. Math. 73(3), 1983: 349–66; ibid. 75(2) (1984), 381]. New York: Springer, 1986: 9–27.

[9] Gerd Faltings. Diophantine approximation on abelian varieties. *Ann. of Math. (2)*, 133(3), 1991: 549–76.

[10] Phillip Griffiths and Joseph Harris. *Principles of algebraic geometry*. Pure and Applied Mathematics. New York: Wiley-Interscience [John Wiley & Sons], 1978.

[11] Robin Hartshorne. *Algebraic geometry*. Vol. 52 of Graduate Texts in Mathematics, New York and Heidelberg: Springer-Verlag, 1977.

[12] Marc Hindry and Joseph H. Silverman. *Diophantine geometry: An introduction*, Vol. 201 of *Graduate Texts in Mathematics*. New York: Springer-Verlag, 2000.

[13] Serge Lang. *Fundamentals of Diophantine geometry*. New York: Springer-Verlag, 1983.

[14] Serge Lang. *Algebraic number theory*, Vol. 110 of Graduate Texts in Mathematics, 2nd ed. New York: Springer-Verlag, 1994.

[15] Serge Lang. *Algebra*, Vol. 211 of Graduate Texts in Mathematics, 3rd ed. New York: Springer-Verlag, 2002.

[16] Brian Lawrence and Akshay Venkatesh, Diophantine problems and p-adic period mappings. *Invent. Math.*, 221(3), 2020: 893–999.

[17] Hideyuki Matsumura. *Commutative algebra*, Vol. 56 of Mathematics Lecture Note Series, 2nd ed. Reading, MA: Benjamin/Cummings Publishing Co., Inc., 1980.

[18] Hideyuki Matsumura. *Commutative ring theory*, Vol. 8 of Cambridge Studies in Advanced Mathematics, 2nd ed. Trans. M. Reid. Cambridge: Cambridge University Press, 1989.

[19] Louis J. Mordell. On the rational solutions of the indeterminate equations of the third and fourth degrees. *Proceedings of the Cambridge Philosophical Society*, 21, 1922–23: 179.

[20] Atsushi Moriwaki. *Arakelov geometry*, Vol. 244 of *Translations of Mathematical Monographs*. Trans. 2008. Providence, RI: American Mathematical Society, 2014.

[21] David Mumford. *Lectures on curves on an algebraic surface*. With a section by G. M. Bergman. Vol. 59 of Annals of Mathematics Studies. Princeton, NJ: Princeton University Press, 1966.

[22] David Mumford. *Abelian varieties*, Vol. 5 of Tata Institute of Fundamental Research Studies in Mathematics. Corrected reprint of the 2nd (1974) ed. New Delhi: Hindustan Book Agency, 2008.

[23] Jürgen Neukirch. *Algebraic number theory*, Vol. 322 of Grundlehren der mathematischen Wissenschaften [Fundamental Principles of Mathematical Sciences]. Trans. Norbert Schappacher. Berlin: Springer-Verlag, 1999.

[24] Alexander Polishchuk. *Abelian varieties, theta functions and the Fourier transform*, Vol. 153 of *Cambridge Tracts in Mathematics*. Cambridge: Cambridge University Press, 2003.

[25] Paulo Ribenboim. Recent results about Fermat's last theorem. *Exposition. Math.*, 5(1), 1987: 75–90.

[26] Paulo Ribenboim. *Fermat's last theorem for amateurs*. New York: Springer-Verlag, 1999.

[27] Wolfgang M. Schmidt. *Approximation to algebraic numbers*, Vol. 19 of l'Enseignement Mathmatique, Série des Conférences de l'Union Mathématique Internationale, No. 2. Geneva: Secrétariat de l'Enseignement Mathématique, Université de Genève, 1972.

[28] Wolfgang M. Schmidt. *Diophantine approximation*, Vol. 785 of *Lecture Notes in Mathematics*. Berlin: Springer, 1980.

[29] Paul Vojta. Siegel's theorem in the compact case. *Ann. of Math. (2)*, 133(3), 1991: 509–48.

[30] Paul Vojta. Applications of arithmetic algebraic geometry to Diophantine approximations. In Edoardo Ballico (ed.), *Arithmetic algebraic geometry: Trento, Italy*

1991, Vol. 1553 of Lecture Notes in Mathematics, pp. 164–208. Berlin: Springer, 1993.

[31] André Weil. *Variétés abéliennes et courbes algébriques*, Vol. 1064 of Actualités Sci. Ind., Publ. Inst. Math. Univ. Strasbourg 8 (1946). Paris: Hermann & Cie., 1948.

Notation

N 1 The integers, rational numbers, real numbers, and complex numbers are respectively denoted by $\mathbb{Z}$, $\mathbb{Q}$, $\mathbb{R}$, and $\mathbb{C}$. The nonnegative integers, nonnegative rational numbers, and nonnegative real numbers are respectively denoted by $\mathbb{Z}_{\geq 0}$, $\mathbb{Q}_{\geq 0}$, and $\mathbb{R}_{\geq 0}$.

N 2 A ring is assumed to be commutative with identity. In particular, so is a field. If A is a ring, then $A^{\times}$ denotes the set of units of A. The set of $m \times n$ matrices with entries in A is denoted by $M(m,n;A)$. The group of $n \times n$ invertible matrices with entries in A is denoted by $\mathrm{GL}(n,A)$.

N 3 For integers a,b, we write $a \mid b$ if $b = ca$ for some integer c. Given integers $a_1, \ldots, a_n$, the greatest common divisor of $a_1, \ldots, a_n$, denoted by $\mathrm{GCD}(a_1, \ldots, a_n)$, is the nonnegative integer d with $a_1 \mathbb{Z} + \cdots + a_n \mathbb{Z} = d\mathbb{Z}$.

N 4 We declare that the degree of the zero polynomial is equal to $-\infty$. If f and g are polynomials with coefficients in an integral domain, then we have

$$\deg(f+g) \leq \max\{\deg(f), \deg(g)\}, \qquad \deg(fg) = \deg(f) + \deg(g).$$

N 5 Let f be a polynomial in $X_1, \ldots, X_n$ with coefficients in a ring. Given distinct variables $X_{i_1}, \ldots, X_{i_e}$ among $X_1, \ldots, X_n$, we denote by $\deg_{(X_{i_1}, \ldots, X_{i_e})}(f)$ the degree of f with respect to $X_{i_1}, \ldots, X_{i_e}$. If $e = 1$, we often denote $\deg_{(X_i)}(f)$ by $\deg_{X_i}(f)$ or by $\deg_i(f)$ to simplify the notation.

N 6 Let f be a polynomial in $X_1, \ldots, X_n, Y_1, \ldots, Y_m$ with coefficients in a ring A. We say that f is a polynomial of bidegree at most (a,b) if

$$\deg_{(X_1, \ldots, X_n)}(f) \leq a \quad \text{and} \quad \deg_{(Y_1, \ldots, Y_m)}(f) \leq b.$$

Further, if f is expressed as a sum of the following form:

$$cX_1^{a_1} \cdots X_n^{a_n} Y_1^{b_1} \cdots Y_m^{b_m} \qquad (a_1 + \cdots + a_n = a, \ b_1 + \cdots + b_m = b, \ c \in A),$$

163

then f is called a bihomogeneous polynomial of bidegree (a, b). Bihomogeneous polynomials of bidegree (a, b) are characterized as polynomials that satisfy

$$f(\lambda X_1, \ldots, \lambda X_n, \mu Y_1, \ldots, \mu Y_m) = \lambda^a \mu^b f(X_1, \ldots, X_n, Y_1, \ldots, Y_m)$$

for new variables λ and μ.

We remark that a homogeneous polynomial with

$$\deg_{(X_1, \ldots, X_n)}(f) = a \quad \text{and} \quad \deg_{(Y_1, \ldots, Y_m)}(f) = b$$

need not be a bihomogeneous polynomial of bidegree (a, b). For example, the polynomial $f = X + Y$ is a homogeneous polynomial with $\deg_X(f) = \deg_Y(f) = 1$, but f is not a bihomogeneous polynomial of bidegree $(1, 1)$.

N7   Let X be an algebraic scheme over a field F (i.e., a scheme of finite type over F). For a point x of X, we denote by $\kappa(x)$ the residue field at x. For an F-algebra R, the set of all R-valued points of X is denoted by

$$X(R) = \{\text{morphism } \mathrm{Spec}(R) \to X \text{ over } R\}.$$

An element of $X(F)$ is called a rational point of X over F, or an F-rational point for short. An algebraic variety is an algebraic scheme that is irreducible and reduced.

N8   An algebraic variety of dimension one is called an algebraic curve, or a curve for short. An algebraic variety of dimension two is called an algebraic surface, or a surface for short.

N9   Let X be an algebraic variety, and let ϕ be a nonzero rational function on X. The principal divisor associated to ϕ is denoted by (ϕ). Two Cartier divisors D, D' on X are said to be linearly equivalent, which we denote by $D \sim D'$, if $D - D' = (\phi)$ for some nonzero rational function ϕ on X.

N 10   Let X be an algebraic variety. The Picard group of X, denoted by $\mathrm{Pic}(X)$, is an abelian group consisting of all isomorphim classes of line bundles on X with binary operation given by the tensor product $\otimes$.

N11   Let

$$\begin{cases} X = \mathbb{P}^n = \mathrm{Proj}(F[X_0, \ldots, X_n]), \\ Y = \mathbb{P}^m = \mathrm{Proj}(F[Y_0, \ldots, Y_m]) \end{cases}$$

be projective spaces over a field F. We set $Z = X \times Y$. Let $p_1 : Z \to X$ and $p_2 : Z \to Y$ be the natural projections. For $(a, b) \in \mathbb{Z} \times \mathbb{Z}$, we set $\mathcal{O}_Z(a, b) = p_1^*(\mathcal{O}_X(a)) \otimes p_2^*(\mathcal{O}_Y(b))$. Then the map $\mathbb{Z} \times \mathbb{Z} \to \mathrm{Pic}(Z)$ given by $(a,b) \mapsto \mathcal{O}_Z(a,b)$ is an isomorphism of abelian groups (see Remark 3.20). In particular, if D is a divisor on Z, then there is a unique (a,b) with $\mathcal{O}_Z(D) \cong \mathcal{O}_Z(a,b)$. In this case, D is called a divisor of bidegree (a,b).

Suppose that a,b are nonnegative integers. Then a nonzero element of $H^0(Z, \mathcal{O}_Z(a,b))$ is naturally regarded as a bihomogeneous polynomial of bidegree (a,b) with respect to $X_0, \ldots, X_n$ and $Y_0, \ldots, Y_m$.

Let D be an effective divisor on Z, and we write $\mathcal{O}_Z(D) \cong \mathcal{O}_Z(a,b)$ as above. Then a,b are nonnegative integers, and D is defined by a bihomogeneous polynomial of bidegree (a,b).

Index of Symbols

$\deg_i(F)$, 79, 163
Δ, 62, 123, 130
Δ', 66, 123, 130
$D_{K/\mathbb{Q}}$, 10
$h(F)$, 84
$\hat{h}_L(x)$, 55
$h_K(x)$, 29
$h_L(x)$, 36
$h_\phi(x)$, 35
$h^+(F)$, 84
$h_K^+(x)$, 29
$h^+(x)$, 30
$h(s)$, 132
$h(x)$, 30
$|I|_d$, 85
I, 88
$\mathrm{ind}_x(f;d)$, 85
$\mathrm{ind}_a(P;d)$, 87
J, 48, 61, 120
j, 121
$K(\mathbb{C})$, 9, 74
$\langle x, y \rangle$, 121
$\langle x, y \rangle_L$, 56

$\log^+(a)$, 29
M_K, 27
$[n]$, 48, 49, 56, 70
$\mathrm{Norm}_\pi(D)$, 98
$\mathrm{Norm}_\pi(L)$, 98
$\mathrm{ord}_P(x)$, 26
$\partial^{a,b}$, 89
∂_I, 87
$j_a(x)$, 62
$\Phi_{N\theta}$, 129
$\Phi_{N\theta}^{ij}$, 129
$\Phi_{N\theta}^{ijk}$, 129
Θ, 63
θ, 120, 129
$V(d_1, d_2, d)$, 124, 130, 131
$|F|_\infty$, 79
$|F|_v$, 77
$|I|$, 34, 88
$|x|_v$, 26, 27
$|x|$, 121
$\|x\|_K$, 74
X^I, 34, 77
$x \otimes y$, 31

Index

Abel–Jacobi map, 60, 62
  injectivity of —s, 62, 121
absolute value
  archimedean —, 68
algebraic integer, 9
algebraic scheme, 164

bidegree
  — of a polynomial, 101–5, 107, 109, 110,
    131, 135, 137, 138, 140, 141, 143–45,
    147, 163, 164

centrally symmetric subset, 15
convex body, 15
curve, 164
  algebraically equivalent —s, 51
  Fermat — of degree n, 156

Dedekind domain, 11, 69
degree
  — of a polynomial, 79, 163
difference, 23
Diophantine geometry, 1, 4, 5, 25, 73, 118
discrete valuation, 13
discrete valuation ring, 12
discriminant, 8
  — over $\mathbb{Q}$, 10

F-rational point, 112, 120, 164
Faltings's theorem, 73, 117–21, 129, 156, 157
Fermat property, 157
field of definition, 41

formula
  Jensen's —, 81
  product —, 27, 31, 68, 151
  projection —, 51, 133, 144
fractional ideal (of a ring of integers), 12

Gauss's lemma, 78, 100, 103, 105, 109
Gelfond's inequality, 84
genus (of a curve), 117, 157
Grauert's theorem, 46

height
  absolute (logarithmic) Weil —, 29, 32, 42,
    84, 149, 156
  canonical —, 55
  — associated to a line bundle, 36, 38, 149
  — of a global section of a Vojta divisor,
    124, 125, 132
  (logarithmic) — of a point in projective
    space, 30
  — of a polynomial, 84, 90, 99, 107, 144
  (logarithmic) — of a vector, 29–31
  naive —, 32
  Néron–Tate —, 55
  Néron–Tate — pairing, 56, 66, 72, 121
  positivity of —s, 38
Hermite–Minkowski theorem, 19, 67, 69
homomorphism (of group varieties), 44, 45, 55

index
  — of a formal power series, 85
  — of a global section, 124, 125, 143, 146,
    149, 150
  — of a polynomial, 87

inequality
  Gelfond's —, 84
  Vojta's —, 121
inner product, 14
integral basis, 10, 18

Jensen's formula, 81

Kronecker's theorem, 42

lattice, 14, 18, 28
  volume of a —, 15
Leibniz rule, 114, 116, 150
lemma
  Gauss's —, 78, 100, 103, 105, 109
  rigidity —, 44
  Roth's —, 88, 89, 91, 92, 143, 145, 146
  Siegel's — (for $\mathbb{Z}$), 74
  Siegel's — (for algebraic integers), 76, 141
length
  — of a polynomial with respect to v, 77, 90
line bundle
  even — (on an abelian variety), 49, 55, 56, 63, 66
  odd — (on an abelian variety), 49, 55
  Poincaré —, 61, 65
local Eisenstein theorem, 112, 146, 153

Mahler measure, 79
Minkowski's convex body theorem, 16, 58
Minkowski's discriminant theorem, 19
Mordell–Weil theorem, 58, 67, 71, 121

norm
  — (of an element of an extension field), 6
  — of a Cartier divisor, 98
  — of an invertible sheaf, 95, 96, 98
Northcott's finiteness theorem, 39, 41, 42, 56, 58, 68, 72, 121, 123
number field, 9

Picard group, 164
place
  archimedean —, 79, 84, 101, 103–5, 109, 110, 141, 143, 147, 148, 152, 153
  nonarchimedean —, 31, 78, 84, 100, 103, 105, 109, 143, 147, 148, 151, 153
polynomial
  bihomogeneous —, 107, 131, 135–38, 141, 144, 164

product formula, 27, 68, 151
projection formula, 51, 133, 144

R-valued point, 9, 164
ramification index, 22
ramified, 22–24, 69
residue degree, 22, 24
Riemann–Roch theorem, 133, 134
rigidity lemma, 44
ring of integers, 9, 11–13
Roth's lemma, 88, 89, 91, 92, 143, 145, 146

seesaw theorem, 45
Serre's duality theorem, 133
Siegel's lemma (for $\mathbb{Z}$), 74
Siegel's lemma (for algebraic integers), 76, 141

theorem
  Faltings's —, 73, 117–21, 129, 156, 157
  Grauert's —, 46
  Hermite–Minkowski —, 19, 67, 69
  Kronecker's —, 42
  local Eisenstein —, 112, 146, 153
  Minkowski's convex body —, 16, 58
  Minkowski's discriminant —, 19
  Mordell–Weil —, 25, 58, 67, 71, 121
  Northcott's finiteness —, 39, 41, 42, 56, 58, 68, 72, 121, 123
  Riemann–Roch —, 133, 134
  seesaw —, 45
  Serre's duality —, 133
  — of the cube, 47
  — of the square, 50
  weak Mordell–Weil —, 70, 72
theorem of the cube, 47
theorem of the square, 50
theta divisor, 60, 63, 66, 121
torsion point, 56
trace (of an element of an extension field), 6
  — form, 8

unramified, 22–24, 69

variety
  abelian —, 43–45, 47, 49–52, 54–56, 61, 69–71

algebraic —, 164
geometrically irreducible algebraic —, 43
group —, 43, 44
Jacobian —, 48, 59–61, 120
Picard —, 61
Vojta divisor, 124, 126, 127, 130, 131, 133

sequence of polynomials expressing a
    global section of a —, 132
Vojta's inequality, 121

weak Mordell–Weil theorem, 70, 72
Wronskian of a polynomial, 88